GOD'S DESIGN FOR LIFE

The Animal Kingdom

Debbie Lawrence and Richard Lawrence

R and D Educational Center
Shining the Light on God's Design

God's Design for Life is a complete life science curriculum for elementary aged children. The books in this series are designed for use in the Christian homeschool, and provide easy to use lessons that will encourage children to see God's hand in everything around them.

Published by
R and D Educational Center, LLC
301 Immigrant Trail
Windsor, CO 80550

You may contact the authors at:
info@rdeducation.com
(970) 686-5744

Visit our website at
www.rdeducation.com

Contents

Welcome to *God's Design for Life*

God's Design for Life is a series that has been designed for use in teaching life science to elementary aged children. It is divided into three books: *The World of Plants, The Animal Kingdom,* and *The Human Body.* Each book has approximately 30-35 lessons as well as a unit project that ties all of the lessons together.

In addition to the lessons, special features in each book include biographical information on several scientists involved in the development of biology as we know it today as well as interesting facts pages to make the subject more fun and a little less dusty.

Please use the books in this series as a guide and feel free to add to each lesson. The information included here is just a beginning. A resource guide is included in Appendix B to help you find additional information and resources. Also, a supply list of items needed is included at the beginning of each lesson. A master list of all supplies needed for the entire book can be found in Appendix C.

If you wish to cover the material in this series in one year you should plan on covering approximately 3 lessons per week. The time required for each lesson varies depending on how much additional information you want to include, but you can plan on about 45 minutes per lesson. Older children can do quizzes, tests or additional activities on non-lesson days if you choose to do science every day.

If you wish to cover material in more depth you can add additional information and take a longer period of time to cover all the material or you could choose to do only one or two of the books in the series as a unit study.

Why Teach Life Science?

Maybe you hate science or you just hate teaching it. Maybe you love science but don't quite know how to teach it to your children. Maybe science just doesn't seem as important as some of those other subjects you need to teach. Maybe you need a little motivation. If any of these descriptions fits you, then please consider the following.

It is not uncommon to question the need to teach your kids hands-on science in elementary school. We could argue that the knowledge gained in science will be needed later in life in order for your children to be more productive and well-rounded adults. We could argue that teaching your children science also teaches them logical and inductive thinking and reasoning skills, which are tools they will need to be more successful. We could argue that science is a necessity in this technological world in which we live. While all of these arguments are true, none of them are the real reason that we should teach our children science. The most important reason to teach science in elementary school is to show your children how God is in everything we see, hear, and do.

God is the Master Creator of everything. He is all around us. All of the laws of physics, biology, and chemistry were put in place by the Great Creator. These laws were put here for us to see His wisdom and power. In science we see the hand of God at work more than in any other subject. Romans 1:20 says, "From the time the world was created, people have seen the earth and sky and all that God made. They can clearly see his invisible qualities – his eternal power and divine nature. So they have no excuse whatsoever for not knowing God." (NLT). We need to help our children see God in the world around them so they will be able to recognize God and follow Him.

The study of life science helps us understand the balance of nature so that we can be good stewards of our bodies, the plants, and the animals around us. It helps us appreciate the intricacies of life and the wonder of God's creation. Understanding the world of living things from a biblical point of view will prepare our children to deal with an ecology obsessed world. Realizing that science does not contradict what God says in the Bible will help our children defend their faith. We need to teach science to our children so they will realize that everything the Bible says is true.

It's fun to teach life science! It's interesting too. Children have a natural curiosity about living things so you won't have to coax them to explore the world of living creatures. You just have to direct their curiosity and reveal it to them.

Finally, teaching life science is easy. It's all around us. Everywhere we go, we are surrounded by living things. You won't have to try to find strange materials for experiments or do dangerous things to learn about life.

How Do I Teach Science?

In order to teach any subject you need to understand how people learn. People learn in different ways. Most people, and children in particular, have a dominant or preferred learning modality in which they absorb and retain information more easily.

If your child's dominant modality is:

Auditory – he needs to not only hear the information but he needs to hear himself say it. This child needs oral presentation as well as oral drill and repetition.

Visual – he needs things he can see. This child responds well to flashcards, pictures, charts, models, etc.

Kinesthetic – he needs active participation. This child remembers best through games, hands-on activities, experiments, and field trips.

Also, some people are more relational and others are more analytical. Your relational child needs to know who the people are, why this is important, and how it will affect him personally, whereas your analytical child wants just the facts.

If you are trying to teach more than one child you will probably have to deal with more than one learning style. Therefore, you need to present your lessons in several different ways so that each child can grasp and retain the information. You need to give them a reason to learn it.

To help you with this, we have divided each lesson into three sections. The first section introduces the topic. It is the "just the facts" part of the lesson for the analytical child. This section is marked by the 𝕊 icon. The second section is the observation and hands-on section denoted by the 👁 ✋ icon. This section helps your visual and kinesthetic learners. The final section is the summary and review section denoted by the 🎁 icon representing wrapping up the lesson. This oral review helps your auditory learners. Also included here is the applications part of the lesson to help your relational child appreciate what he has learned. We have included periodic biographies to help your child appreciate the great men and women who have gone before us in the field of science.

We suggest a threefold approach to each lesson:

𝕊 Introduce the topic

- We give a brief description of the facts. Frequently you will want to add more information than the bare essentials given in this book. This section of each lesson is written as if we were talking to your child. In addition to reading this section aloud, you may wish to do one or more of the following:

- Read a related book with your child.
- Write things on the board to help your visual child.
- Give some history of the subject. We provide some historical sketches to help you, but you may want to add more.
- Ask questions to get your child thinking about the subject.

👁 ✋ Make observations and do experiments

- One or more hands-on projects are suggested for each lesson. This section of each lesson is written to the parent/teacher.
- Have your child observe the topic for him/herself whenever possible.

🎁 Wrap it up

- The "What did we learn?" section has review questions.
- The "Taking it further" section encourages your child to
    o Draw conclusions
    o Make applications of what was learned
    o Add extended information to what was covered in the lesson
- The "FUN FACT" section adds fun information.
(Questions with answers are provided to help you wrap up the lesson.)

By teaching all three parts of the lesson you will be presenting the material in a way that all learning styles can both relate to and remember.

Also, this method relates directly to the scientific method and will help your child think more scientifically. Don't panic! The "scientific method" is just a way to logically examine a subject and learn from it. Briefly, the steps of the scientific method are:
1. Learn about a topic.
2. Ask a question.
3. Make an hypothesis (a good guess).
4. Design an experiment to test your hypothesis.
5. Observe the experiment and collect data.
6. Draw conclusions. (Does the data support your hypothesis?)

Note: It's okay to have a "wrong hypothesis." That's how we learn. Be sure to try to understand why you got a different result than you expected.

Our lessons will help your child begin to approach problems in a logical, scientific way.

How Do I Teach Creation vs. Evolution?

We are constantly bombarded by evolutionist ideas about living things. Did dinosaurs really live millions of years ago? Did man evolve from apes? Which came first, Adam and Eve or the cavemen? Where did living things come from? How can we teach our children the truth about the origins of life? The Bible answers these questions and this book takes a literal interpretation of the Bible. We believe this is the only way we can teach our children to trust that everything God says is true.

There are 5 common views of the origins of life:

1. Literal biblical account – Each day in Genesis is a literal day of creation in which God created everything that exists. (Genesis is literal and true.)
2. Day/age theory – Each day in Genesis is really a long period of time during which evolution occurred. (Genesis is a figurative outline.)
3. Gap theory – Evolution occurred in a supposed long period of time between Genesis 1:1 and Genesis 1:2 (Genesis is incomplete.)
4. Theistic evolution – God guided the evolutionary process over billions of years. (Genesis is a fable.)
5. Evolution – there is no God, everything happened by chance over billions of years. (Genesis is a fable and there is no God.)[1]

Any theory that combines evolution and creation presupposes that death entered the world before Adam sinned which contradicts what God has said. Genesis must be taken literally to avoid compromising what God says is true in His word.

Viewed objectively, the evidence clearly indicates an intelligent creator not chance evolution. The volume of evidence supporting creation is massive and cannot be adequately covered in this book. If you would like more information on this topic please see the resource guide in Appendix B. To help get you started, just a few examples of evidence supporting creation are given below:

→ **Evolutionary Myth**: There has been some kind of life on earth for billions of years. And man has lived for one million years. **The Truth**: If animals and people have lived for millions of years there would be trillions of people and animals on the earth today, even if we allowed for worst-case plagues, natural disasters, etc. The number of people on earth today is about six billion. Even allowing for lower birth rates and higher death rates than what we have traditionally seen, this number indicates that man has been around for only a few thousand years.[2]

---

[1] Ken Ham et al., *The Answers Book*, (El Cajon: Master Books, 1992), 89-101.
[2] John D. Morris, Ph.D., *The Young Earth*, (Colorado Springs: Creation Life Publishers, 1994), 70-71.

→ **Evolutionary Myth**: Man evolved from an ape-like creature. **The Truth**: All "missing links" showing human evolution from apes have been proven to be mistaken identity or deliberate hoaxes. The links remain missing.[3]

→ **Evolutionary Myth**: All animals evolved from lower life forms. Darwin stated that transitional fossils would be discovered to show the chain of evolution. **The Truth**: No transitional fossils have been found showing one species or type of animal changing or evolving into another. For example, there are no fossils showing something that is part way between a dinosaur and a bird.[4]

→ **Evolutionary Myth**: Transitional forms survived better than the original species. **The Truth:** A transitional form of an animal, such as one in which an arm slowly changes into a wing, would be very vulnerable because it could not fight and it could not fly away. This would make it less likely to survive not more likely.[5]

→ **Evolutionary Myth**: Dinosaurs evolved into birds. **The Truth**: The differences between reptile and bird respiratory systems are substantial. It is impossible to believe that an animal could make that many changes over time and still survive.[6]

→ **Evolutionary Myth:** Thousands of chance changes over millions of years resulted in the living creatures we see today. **The Truth:** What is now known about human and animal anatomy shows the body structures to be infinitely more complex than was believed when Darwin published his work. Many biologists and especially microbiologists are now saying that there is no way these complex structures could have developed by accident.[7]

Since the evidence does not support their theories, evolutionists are constantly coming up with new ways to try to support what they believe. One of the newer ideas is the jump theory, which says that a dinosaur laid an egg and a bird hatched out of it. That is completely contradictory to everything we observe and is not science at all. It is wishful thinking. We need to teach our children the difference between science and wishful thinking.

You can be confident that if you teach that God created the world and that what the Bible says is true, that the evidence will support what you teach. Instill in your child a confidence in the truth of the Bible. If scientific thought seems to contradict the Bible, realize that scientists are human and make mistakes, but God does not lie. At one time scientists believed that the earth was the center of the universe, that the earth was flat, that man could not travel more than 37 miles per hour, and that blood letting was good for the body. All of these were believed to be scientific facts but have since been disproved, but the Word of God remains true.

---

[3] Duane T. Gish, Ph.D. *The Amazing Story of Creation from Science and the Bible*, (El Cajon: Institute for Creation Research, 1990), 78-83.

[4] Ibid., 36, 53-60.

[5] Ibid., 46.

[6] Gregory Parker et. al., *Biology God's Living Creation*, (Pensacola: A Beka Books, 1997) 474-475.

[7] Ibid. 384-385.

Lesson

1

The Animal Kingdom

Is it a mouse or a moose?

Supply list:

- Just your imagination

Note: *The Usborne Illustrated Encyclopedia of the Natural World* is an excellent resource for additional information on the animal kingdom and is available at our web site.

Animals and plants are the two largest and most familiar groups of living things. The most distinguishing difference between plants and animals is that plants can make their own food and animals cannot. Animals must eat plants or other animals to obtain energy. Because of this, animals are mobile. They can move about in order to obtain food.

Animals come in all shapes and sizes. Some are so tiny you can only see them with a microscope. Others are as huge as a car or even a house. There are over 1 million different species of animals that have been identified and classified and millions more that have not been classified.

In order to study so many different types of animals it is convenient to group them together by their similar characteristics. The first grouping that scientists make is to divide animals by whether they have a backbone or not. Animals with a backbone are called vertebrates. Animals without a backbone are called invertebrates.

Although only 3% of all animals are vertebrates, they are the animals we are most familiar with. Vertebrates are the animals we see around us every day. Every vertebrate has a backbone with a spinal cord through it ending in a brain. Vertebrates have the same major systems that humans have, including skin, skeletal, muscular, nervous, respiratory, and

1

digestive systems. Although all of these systems occur in all vertebrates, they vary considerably among the different species.

Vertebrates are divided into five different groups: mammals, birds, amphibians, reptiles, and fish. We will explore each of these groups in more detail.

Invertebrates are animals without spinal cords. They are very diverse and account for nearly 97% of all animals. Invertebrates do not have internal skeletons. Invertebrates include sponges, jellyfish, worms, insects, and many more creatures. We will also study each group of invertebrates in more detail.

Animal Charades:

Have your child pretend to be an animal and have everyone else guess what animal he is. Whoever guesses the animal correctly gets to be the next animal. Encourage your children, especially older children, to choose animals other than mammals, with which they are most familiar.

What did we learn?

What are the two major divisions of animals? (Vertebrates and invertebrates)

What similarities are there among all animals? (They move, must eat plants or other animals)

Taking it further

When did God create the animals? (On the 5th day He created fish and birds, on the 6th day He created land animals. See Genesis chapter 1.)

How is man different from animals? (Humans have a conscience so they can tell right from wrong. Animals act on instinct. People have souls so they can know God. People were made in God's image to have a relationship with Him. Despite our physical similarities, people are spiritually very different from animals. God gave man dominion over the animals. See Genesis 1:28.)

Vertebrates

Does it have a backbone?

Supply list:

- 3 ring binder

- 12 or more dividers with tabs

The animals we are most familiar with are vertebrates. Vertebrates are animals that have a spinal cord inside a backbone that ends in a brain. Vertebrates can be classified into five categories: mammals, birds, amphibians, reptiles, and fish. These are the animals we notice most around us because, in general, they are the largest animals. Although each of these groups of animals has unique characteristics, they have some common characteristics as well.

All vertebrates have a spinal cord and a brain. These are the major parts of each vertebrate's nervous system. The spinal cord is protected by a backbone, which is really a series of smaller bones called vertebrae, hence the name vertebrates. Messages travel from the animal's brain down the spinal cord to the various parts of the body to tell the animal how to move and what to do. Messages also travel from the various parts of the body along the spinal cord to the animal's brain. Vertebrates have some of the most complex nervous systems of all the animals.

Another common trait that is unique to vertebrates is an internal skeleton. This skeleton is what allows vertebrates to be much larger than most other animals. God gave vertebrates the internal structure needed to support the weight of a large body. Not all vertebrates are large, but nearly all large animals are vertebrates. A few exceptions are the octopus and giant nautilus. These creatures can be large without an internal skeleton because the water in which they live helps to support their weight. Also, for the most part, vertebrates have more complex muscular, digestive, and respiratory systems than invertebrates.

We will discuss each group of vertebrates in more detail in the following lessons, but here is a quick overview of the major types of vertebrates. Mammals are vertebrates with hair or fur. They are warm-blooded and they nurse their young. Birds are warm-blooded animals with feathers. The other vertebrates are all cold-blooded animals. Amphibians are unique because they begin life in the water and as they mature their bodies change and they begin to breathe air through lungs. Reptiles are animals with scales that breathe air. And fish are aquatic animals that have gills that extract oxygen from the water in which they live. Vertebrates are easy to find and fun to study. Enjoy learning more about God's wonderful creatures.

Animal Kingdom Notebook:

As you study the animal kingdom your child will be making a notebook that will include his/her projects. Today have your child start his/her notebook by making dividers for each part of the animal kingdom. Use the dividers with tabs that are designed for three ring binders. Have your child make labels for each tab in the notebook. Tabs need to be labeled as follows:

Mammals, Birds, Amphibians, Reptiles, Fish, Arthropods, Mollusks, Coelenterates, Echinoderms, Sponges, Worms, Protists, and Monerans (You may combine Protists and Monerans if you wish.)

Explain that these are the various parts of the animal kingdom that you will be studying. Have your child name as many animals in each category above as possible. Some, like mammals, will be very easy, but your child may have no idea what animals belong in some of the other categories. You can include anything in your child's notebook that you wish. Some ideas include: the projects from this book, pictures of projects or activities that you do, pictures from field trips, photos cut from magazines or coloring books, and drawings. Use your imagination.

What did we learn?

What are the two major divisions of the animal kingdom? (Vertebrates and invertebrates)

What characteristics define an animal as a vertebrate? (Vertebrates have a spinal cord ending in a brain. They also have internal skeletons.)

What are the five groups of vertebrates? (Mammals, birds, amphibians, reptiles, and fish)

Taking it further

Think about pictures you have seen of dinosaur skeletons. Do you think dinosaurs were vertebrates or invertebrates? Why do you think that? (Dinosaurs were vertebrates. This is shown by the fact that dinosaurs have internal skeletons and these skeletons contain vertebrae along the backs of the animals.)

Mammals

The fuzzy creatures

Supply list:

- 1 copy of "Mammals Have Fur" worksheet per child (pg. 7)

- Samples of hair from as many mammals as possible

- Book showing pictures of different mammals

The most familiar vertebrates on earth are mammals. How can you tell if an animal is a mammal? Mammals have five common characteristics. They are warm-blooded, they have hair, they give live birth, they feed milk to their young, and they breathe air through lungs. There is great variety among mammals. Some are tiny like mice; others are very large like the giraffe and the elephant. Most live on land but a few, dolphins and whales for example, live in the water. To identify an animal as a mammal however, we must examine the similarities.

First, mammals are warm-blooded. This means that their bodies stay about the same temperature regardless of the temperature of the air around them. A mammal's body regulates or controls its body temperature. To produce heat for the body, mammals must eat a lot of food.

Second, most mammals give birth to live young. Two exceptions are the spiny anteater (echidna) and the platypus, both of which lay eggs. Yet, even these animals feed their young milk from special glands in their bodies. These glands are called mammary glands, hence the name mammal. The major deciding factor in an animal being a mammal is whether it nurses its young or not.

In addition to these common characteristics, all mammals have hair or fur on their bodies. Some mammals seem completely covered with hair while others have just a little hair. Most hair provides protection from the cold. Hair also helps with the sense of touch. And the color and pattern of hair helps many mammals hide from their enemies.

Finally, mammals breathe air through lungs. Even whales and dolphins have lungs and they must surface periodically to get a breath of fresh air, unlike fish that get oxygen from the water itself.

After looking at mammal characteristics you might wonder if humans are mammals. Physically humans are mammals. However, we know that man is not an ordinary animal. Man is a spiritual and moral creature who can have a relationship with God. Man was created in God's image.

Mammals Have Fur:

Help your child complete the "Mammals Have Fur" worksheet. Use pictures of mammals to describe the fur on mammals that you do not have access to. Although people are not strictly animals, include a sample of your child's hair and compare it to the hair from other mammals. Add the worksheet to your animal kingdom notebook.

What did we learn?

What five characteristics are common to all mammals? (They are warm-blooded, breathe with lungs, give live birth, nurse their young, and have hair or fur.)

Why do mammals have hair? (To keep them warm, to aid in the sense of touch, and for some it provides camouflage from predators.)

Why is a platypus considered a mammal even though it lays eggs? (It nurses its young.)

Taking it further

Name some ways that mammals regulate their body temperature. (Mammals cool down by sweating or panting. They heat up by eating, exercising, or covering their bodies to keep warm.)

What are some animals that have hair that helps them hide from their enemies? (Tigers and zebras have stripes that make them hard to see. Lions are the color of their surroundings.)

Mammals Have Fur

Animal Name	Hair Length	Hair Color	Parts of body covered	Texture of hair	Sample of hair (If available)
Dog					
Cat					
Pig					
Sheep					
Tiger					
Elephant					
Rabbit					
Armadillo					
Musk-ox					
You (human)					

Mammals Large and Small

Armadillo to Zebra

Supply list:

- Drawing paper

- Markers, colored pencils, or paint

The variety of mammals is astounding. God created thousands of different kinds of mammals. Mammals live in nearly every part of the world including the oceans. We cannot possibly cover every kind of mammal in this book, but we will look at a few of the interesting ones.

Among the largest land mammals are the elephant, the giraffe, and the brown bear. Elephants are the largest land mammals. Adult elephants can weigh as much as six tons and stand up to ten feet high at the shoulder. They have only a little hair around their ears and eyes, but they are still mammals. Female elephants, called cows, and baby elephants, called calves, travel in herds. The oldest female is usually the leader of the herd. Male elephants, called bulls, usually travel alone or with other bulls and only join the herd during mating season. Elephants have long trunks which they use for drinking and for putting food into their mouths. They also have long teeth called tusks that can be used to dig for roots and to remove bark from trees. Elephants are very strong and are sometimes trained by people to carry heavy burdens.

Giraffes are the tallest land mammals. They can grow to be nearly 19 feet tall. This allows giraffes to eat leaves from trees that other animals cannot reach. Being so tall also allows giraffes to see long distances so they can watch for danger. Giraffes can also run very quickly, up to 35 miles per hour, for a short period time. Giraffes live in the African savanna or

8

grassland. One of the most fascinating features of a giraffe is its long tongue, which can be up to 21 inches long!

Bears are another interesting group of land mammals. Grizzly bears are a type of brown bear that lives in most of the northern United States and Canada. A grizzly can be up to 8 feet long from head to rear and weigh about 800 pounds. The Alaskan brown bear can be up to 10 feet long and weigh as much as 1700 pounds.  Bears usually live by themselves after they are about two years old. Bears will eat nearly anything. Although most of a bear's diet consists of plants and berries, it will also eat small animals and fish. When the salmon are swimming upstream, many bears will gather at the edges of the rivers to catch and eat the fish. Bears are very active during the spring, summer and fall. But they sleep most of the winter. During the fall, bears eat nearly constantly to store up enough fat to keep them alive during the winter. They also prepare a den where they will be sheltered from the harsh winter weather. Then they will sleep during the cold weather. When spring arrives, a very hungry bear emerges from its den and begins eating again.

In contrast to the large mammals are some very interesting small mammals. The Pika is a small animal that lives on rocky mountain slopes. It grows to be about 8 inches long and is related to rabbits and hares. Other small mammals include mice, voles, hamsters, and gerbils.

Bats are some of the most unusual mammals. These flying mammals might be confused with birds. But a closer examination will show that bats are very different from birds. Bats have hair, not feathers. And what appear to be wings are actually long fingers connected by a membrane that allows bats to fly. Bats are unusual also because they can detect objects by sending out high-pitched sound waves and sensing their reflections, somewhat like sonar.

Most mammals give live birth but two, the echidna (spiny anteater) and the platypus, are the only mammals that lay eggs. These animals are still considered mammals because they have mammary glands and nurse their young. They are warm-blooded and have hair as well.

God has created a wide variety of mammals. Many are cute and cuddly looking. Others are large and ferocious. But they are all part of the amazing animal kingdom.

Investigating the World of Mammals:

Have your child choose a mammal that he/she wants to learn more about. Then have your child draw a picture of that mammal to include in his/her animal notebook.

For older children, have your child research the animal he/she has chosen and write a report that answers as many of the following questions as possible. Include this report in the animal notebook.

1. What is this animal's habitat - where does it live?

2. How large does this animal grow to be?

3. What does this animal eat?

4. What enemies does this animal have?

5. How quickly does this animal reproduce? How many offspring does it have? How long is the mother pregnant? How long does the baby stay with the mother?

6. What other interesting things did you find out about your animal?

Add your picture and report to your animal kingdom notebook.

What did we learn?

What is the largest land mammal? (Elephant)

What is the tallest land mammal? (Giraffe)

What do bears eat? (Nearly anything, but they prefer plants, roots, and berries.)

Taking it further

What do you think is the most fascinating mammal? Why do you think that?

FUN FACTS

You may not know this about mammals:
- Baby elephants sometimes suck their trunks in the same way that humans suck their thumbs.
- Elephants are pregnant longer than any other animal – 22 months.
- Bears can get cavities from eating too much honey.
- A camel can drink 30 gallons of water in 13 minutes.
- The aye-aye is an animal that only lives on the island of Madagascar.
- The capybara is the world's largest rodent and can weigh 110 pounds.

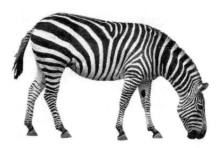

Monkeys and Apes

Primates

Supply list:

- 1 copy of "Mammals Word Search" per child (pg. 14)

One group of mammals that everyone enjoys watching at the zoo is primates. Most people call these animals monkeys, but there are actually three different types of primates. Monkeys are the largest group of primates, but apes and prosimians are also primates. All primates share several common characteristics. First, they have five fingers and five toes. Also, primates have eyes on the front of their faces so they have binocular vision. Many other animals have eyes located more on the sides of their heads and therefore do not have the good depth perception that primates have.

There are 158 different species of monkeys. Some are very small like the pygmy marmoset, which weighs only 8 ounces. The largest monkey is the mandrill, which can weigh as much as 100 pounds. Monkeys live in Central and South America, Africa, and southern Asia. The monkeys that live in the western hemisphere are called New World Monkeys. These monkeys are small to medium sized and have prehensile tails. A prehensile tail is one that is able to grasp onto things and can be used for climbing or swinging. Old World Monkeys live in Africa and Asia. These monkeys are usually larger than the New World Monkeys and do not have prehensile tails.

Monkeys are excellent climbers, using their feet like a second set of hands. Monkeys spend most of their lives in trees and feed on leaves, fruit, flowers, and insects. A few monkeys prey on smaller animals.

The second group of primates is the apes. Apes are very similar to monkeys in appearance with one notable exception; apes do not have tails. Also, apes have arms that are longer than their legs. Common apes include gorillas, chimpanzees, orangutans, and gibbons. Apes live in the tropical forests of Africa and Southeast Asia. Unlike most monkeys, many apes spend a significant amount of time on the ground, although orangutans spend much of their time in the trees.

Gorillas are the largest apes. An adult gorilla weighs about 350 pounds and an adult female weighs about 200 pounds. Gorillas live in groups of 10 or fewer animals with one dominant male, several females, and several young gorillas that are not yet ready to live by themselves. When gorillas are mature they usually leave the group. A male will live by himself until he can find an unattached female to join him and begin a new group. A female will leave and join another group or a lone male.

Chimpanzees are very social apes. They live in groups of at least 12 and up to 100 members. They are mostly herbivorous and eat many different plants. However, they also eat termites and have even been known to eat monkeys and small antelope. Chimps are very creative and use sticks and leaves to help collect termites and water. Other apes, such as orangutans are less social and live more solitary lives.

The third group of primates is the prosimians. At first glance, prosimians may not seem to belong in the same category as apes and monkeys; however, they share the common characteristics of 5 fingers and toes and binocular vision. Prosimians live mostly on the island of Madagascar but some

species live on mainland Africa and in southern Asia. There are 61 species of prosimians including lemurs (shown at the right), galagos, lorises, and bushbabies (shown at the left). Most prosimians have very large eyes. This is helpful for hunting and seeing at night, which is when most prosimians are active.

Mammals Word Search:

Have your child complete the Mammals Word Search. Have your child put the word search in the animal kingdom notebook if you wish.

What did we learn?

What are two common characteristics of all primates? (They have five fingers and five toes, and eyes on the fronts of their heads giving binocular vision.)

What are the three groups of primates? (Monkeys, apes, and prosimians)

What is one difference between apes and monkeys? (Apes do not have tails, monkeys do. Also, apes' arms are longer than their legs, but this is not true for monkeys.)

Where do New World Monkeys live? (In the western hemisphere)

Where do Old World Monkeys live? (In the eastern hemisphere)

What is a prehensile tail? (A tail that can grasp onto things.)

Taking it further

If a monkey lives in South America is it likely to have a prehensile tail? (Yes, because New World Monkeys have prehensile tails and Old World Monkeys do not.)

Are you more likely to find a monkey or an ape in a tree in the rain forest? (You are more likely to find a monkey in a tree. Many apes do not spend a lot of time in trees, whereas most monkeys live the majority of their lives in trees.)

Why do most prosimians have very large eyes? (The majority of prosimians are nocturnal, that is, they sleep during the day and are awake at night. Large eyes allow these animals to see better at night.)

Mammals Word Search

Find the following words in the puzzle below. Words may be horizontal, vertical or diagonal.

Fur	Mammary gland	Warm-blooded	Ape	Monkey
Lemur	Whale	Lungs	Live birth	Zebra
Giraffe	Primate	Bat	Mouse	Camel

```
D G E U T A F R O P K L H Y U
B E N D R M L J I O S C X F A
A Z E B R A Z C V B C A B S R
M N I D S M F V G F T T J A E
S R E I H M U X A A I M U P U
C O L S W A R M B L O O D E D
C A M E L R H O P T W U P E A
K E Y M O Y U S P Z E S R R W
F R L U G G D R W R G E K H U
M L U L E L I V E B I R T H S
O U E N A A T G L A R M D O B
C N H M O N K E Y V A I A N P
F G Y B U D L E W C F A N T L
H S A L E R U P Q Z F L U H E
R A J C J K W H A L E E S O P
```

Man and Monkeys
Did man descend from the apes?

Nearly every library book you pick up about monkeys says that they are relatives of man or that man and monkeys descended from a common ancestor. They point out that humans and apes have many common characteristics so it makes sense that they have common roots. However, the Bible tells a very different story. The Bible says that God made man in His own image (Genesis 1:27). Then God formed woman from man's rib (Genesis 2:22). Humans are a result of God's miracle of creation, not an accident of nature or a series of genetic mutations.

For many years evolutionists have been trying to find the "missing link" between apes and humans. If a fossil of a creature that was partway between a man and an ape could be found it would be very powerful evidence for the theory of evolution. And several claims have been made that the "missing link" has been found. However, when each of these claims has been carefully examined, none have been shown to be something that is half ape and half man. Many examples have been shown to be either just an ape or just a human. Sadly, some examples have been shown to be hoaxes. Let's look at the most famous examples of "missing links".

One of the earliest supposed examples of an ape-man was the Neanderthal Man. In 1860, a few fragments of fossils were discovered in Neanderthal Valley, Germany. Then in 1908, a nearly complete skeleton of a Neanderthal was discovered in France. This skeleton was of a creature with a skull very much like a human but did not walk completely upright. From this discovery many scientists claimed that Neanderthal Man was sub-human. However, after many tests were done on the skeleton it was discovered that it actually belonged to an old man with arthritis. Tests on this skeleton and other bones found in the same area also showed that the people suffered from rickets, a disease that causes bones to become deformed. Also, other Neanderthal skeletons were discovered that indicated the people of that area walked upright and have completely human characteristics. So Neanderthal Man was shown to be just a man.

In 1912, some fossils were discovered near Piltdown, England. These bones included pieces of a jawbone and a skull. Scientists examining the bones declared that the bones were all from one creature that had both human and ape characteristics. This sample was called the Piltdown Man. However, in 1950, scientists declared Piltdown Man to be a hoax. Someone

15

had taken a human skull bone and an ape jawbone and stained them to make them look old. They had also filed the teeth in the jawbone to make them look more human. These "fossils" were then planted in a gravel pit where they were sure to be discovered. Piltdown Man was a hoax.

Nebraska Man was discovered in 1922, in western Nebraska. This discovery consisted of a single tooth, yet scientists declared that it proved the existence of an ape-like man or a man-like ape. The *Illustrated London News* even published a picture of Nebraska Man along with his wife and the tools they used. However, later expeditions unearthed other bones of the supposed ape-man and scientists discovered that the tooth actually belonged to a pig. Nebraska Man was just wishful thinking and bad science.

One of the most famous supposed "missing links" is Lucy. Lucy was discovered by Dr. Donald Johanson in 1973. This skeleton supposedly shows an ape-like creature that walked upright and thus was an ancestor of humans. However, there is great controversy surrounding Lucy. First, the knee bone that supposedly shows Lucy's upright posture was found 200 feet lower and more than 2 miles away from the rest of the skeleton. Second, Lucy's wrist structure has been shown to be consistent with other apes that walk using their knuckles for balance. And third, other skulls that are of the same species as Lucy have been tested and show inner ear characteristics of creatures that do not walk upright. Many scientists now believe that Lucy is the skeleton of an extinct species of ape.

No discoveries have been made that show a direct link between apes and humans. The missing links are still missing and will remain missing because God created both man and apes. So next time you pick up a book that says that man is a descendant of an ape, you can ask where the evidence is for this idea. Because the evidence clearly indicates that apes are apes and humans are humans. You can believe the Bible.

Aquatic Mammals

They live in the water?

Supply list:

- Toothbrush

- Chopped nuts, fruits or vegetables

- Stop watch

When people think of mammals they generally think of furry animals that live on land. They think of monkeys, mice, and tigers. However, not all mammals live on the land. There are several mammals that live in the ocean. These include dolphins, porpoises, and whales. These animals are often thought of as large fish. However, whales, porpoises, and dolphins breathe with lungs and must come to the surface for air on a regular basis. Also, they give birth to live young and nurse their young. Fish cannot do any of these things. In addition, dolphins, porpoises, and whales are warm-blooded and fish are cold-blooded.

Dolphins, porpoises, and whales all have bodies that were designed for living in the water. God gave these mammals sleek bodies that easily glide through the water as well as powerful tail fins called flukes. The fluke moves up and down, instead of side to side like a fish's tail, allowing the animal to dive deep into the water and then resurface quickly for breathing. Because they are designed by God to live in the water, whales, porpoises, and dolphins do not breathe through a nose like most land animals. Instead, they each

have an opening on the tops of their heads called a blowhole through which they breathe. When the animal surfaces, it exhales the air in its lungs causing a spurt of air and a small amount of water to shoot into the sky before the animal takes a new breath.

There are about 90 species of whales, porpoises, and dolphins. Dolphins, porpoises, and many species of whales have teeth. Other species of whales have large comb-like structures in their mouths that they use for straining food from the water. These structures are called baleen and are made from keratin, the same material that your hair and fingernails are made from. Some whales use their baleen to strain out fish and other animals. But the blue whale, which is the largest animal on earth, eats krill, tiny shrimp-like creatures, which are some of the smallest animals on earth. Of course, a blue whale eats about 8,000 pounds of krill each day!

Another mammal that spends it entire life in the water is a manatee. Manatees, which somewhat resemble seals or walruses, live in areas with warm water such as the Florida Everglades and many of the rivers of South America. The manatee, and its relative the dugong, are gentle, slow moving creatures that graze on sea grasses. This grazing habit is often compared to cattle grazing and the manatee is often called a sea cow. Manatees spend most of the time eating and can eat a pound of grass for every ten pounds of weight each day. That means that a 600 pound manatee would eat 60 pounds of plants a day! Like the whales and dolphins, the manatee also has a tail that moves up and down to help it swim and dive. And although the manatee does not have a blowhole, God gave it nostrils on the top of its head so it can surface for air while keeping the majority of its body submerged in water.

God designed most mammals to live on land, but a few were designed to live in the water. So next time you go to the ocean, keep your eyes open for mammals as well as fish.

Acting Like a Whale:

Activity 1: Aquatic mammals live their entire lives in the water, yet they breathe air so they must surface periodically to get a fresh breath. A porpoise can hold its breath for about 4 minutes. Manatees can stay submerged for up to 6 minutes at a time. A bottlenose dolphin can stay underwater for up to 15 minutes. But when it comes to staying submerged, the king of underwater mammals is the sperm whale, which can hold its breath for an hour or more. How long can you hold your breath? Use a stop watch to time how long you can hold your breath.

Activity 2: Baleen whales do not have teeth. Instead they have comb-like ridges that trap food from the water. Chop some nuts, fruits, or vegetables into tiny pieces. Add the pieces to a cup of water. The chopped food represents the tiny creatures that live in the ocean.

Hold a toothbrush sideways over an empty cup and slowly pour the water and food mixture through the bristles of the toothbrush. The bristles will catch some of the food pieces. You can then pull the pieces out of the toothbrush and eat them just like the whale pulls food out of its baleen with its tongue.

What did we learn?

Why are dolphins and whales considered mammals and not fish? (They are warm-blooded, give live birth, nurse their babies, and breathe air with lungs.)

What is the main difference between the tails of fish and the tails of aquatic mammals? (Fish tails move from side to side and mammal tails, or flukes, move up and down.)

What is another name for a manatee? (Sea cow)

Why are manatees sometimes called sea cows? (They move slowly and graze on sea grass and other sea plants just like a cow grazing in a field.)

Taking it further

How has God specially designed aquatic mammals for breathing air? (First, He gave them blowholes or nostrils on the tops of their heads so it is easy to breathe while still being in the water. Second, He designed them to be able to stay submerged for several minutes or even an hour at a time so they do not have to stay near the surface. He also gave them flukes to help them resurface quickly.)

What do you think might be one of the first things a mother whale or dolphin must teach a newborn baby? (One of the first things the mother will do is push the baby toward the surface of the water so it can get its first breath.)

FUN FACT

Some people believe that the legend of mermaids swimming in the ocean may have come from sailors who saw manatees slowly swimming below the surface of the water.

Marsupials

Pouched animals

Supply list:

- Plastic zipper bag	- Glue
- Tag board	- Scissors
- Fake fur or felt	- Construction Paper

Of all the mammals in the world, one of the most entertaining is the kangaroo. Kangaroos hop faster than many animals can run. Kangaroos box each other in a fight for a mate. And kangaroos are often seen, like in the picture above, with the head of a baby poking out of a pouch. These entertaining animals are part of a group of mammals called marsupials.

Marsupials are mammals that give birth to babies that are not fully developed. These tiny babies, depending on the species, can be as small as a grain of rice or as big as a bumble bee. A newly born baby is called a joey and is naked and blind. It uses its sense of smell to crawl along its mother's belly searching for the pouch that will protect it until it is fully developed. Once the joey reaches the pouch, it crawls inside and attaches itself to its mother's mammary gland where it will remain, nursing and growing, for several months.

Kangaroos are the most famous marsupials, but there are many other pouched mammals as well. Koalas, numbats, mulgara, and Tasmanian devils are some of the 258 species of marsupials. Nearly all marsupials live in Australia, Tasmania, and New Zealand. The only marsupial known to live in North America is the opossum.

Red kangaroos are the largest kangaroo and can be 7 to 8 feet tall. They are the largest hopping animals on earth. Yet some breeds of kangaroos are very small. The musky rat kangaroo is only 10-12 inches high. But big or small, all kangaroos have large hind legs with big hind feet. The middle toe of each hind foot is longer than the others and is used for pushing off when hopping. Also, all kangaroos have large tails that help them keep their balance.

Large kangaroos can hop at speeds up to 30 miles per hour. God designed the kangaroo to be a hopping machine. The large legs and tail are ideal. And at the back of each leg is a long stretchy tendon attaching the muscles of the leg to the ankle bones. This tendon stores up energy between hops that is released when the feet hit the ground. When a kangaroo is hopping, its body remains at about the same height, while its legs stretch out and then fold up as it hops. God designed the kangaroo so well for hopping that it uses up about the same amount of energy when it is hopping slowly as when it is hopping quickly.

Kangaroos are generally nocturnal. They spend most of the day sleeping and resting. Then, when the sun goes down and the temperatures cool off, kangaroos begin eating, which they continue doing almost the entire time they are awake. Kangaroos are plant eaters, and like many other plant eaters, they chew their food and swallow it, then later, they spit the food back up and chew it some more. Kangaroos also have special bacteria living in their digestive tracks that eat the cellulose in the plants and help the kangaroos digest the plants.

Koalas, opossums, and kangaroos are all plant eaters. But many other marsupials are meat eaters. The numbat is a marsupial that eats ants and termites. The numbat uses its sharp claws to tear open trees or termite hills. Then it uses its sticky 4-inch long tongue to pick up termites or ants for a tasty meal. A hungry numbat can eat as many as 20,000 termites per day.

The most famous meat-eating marsupial is the Tasmanian devil. Thanks to its sharp teeth and tendency to growl, it has earned a reputation as a very fierce animal. However, recent studies have shown that it is not as fierce as once believed. Tasmanian devils live only on the island of Tasmania near Australia. These animals live in brushy, wooded areas and are nocturnal. They hunt wallabies, wombats, sheep, and rabbits. But they prefer to eat animals that are already dead instead of hunting. Like all marsupials, Tasmanian devils give birth to very tiny young, usually only a fraction of an ounce. The joey then moves to the mother's pouch where it lives for the next 15 weeks.

From numbats to opossums, marsupials are very interesting creatures. See what else you can learn about these pouched animals.

Making a Pouch:

Have your child make his/her own marsupial pouch as follows. On a sheet of tag board or construction paper, draw the belly of a kangaroo. In the center of the sheet glue a plastic zipper bag with the zipper side up. Now glue fake fur or felt across the outside of the zipper bag. Be sure that the fur completely covers the sides and bottom of the zipper bag, but does not block the top of the bag. You now have a pouch. If you have enough fur, you can glue it on the rest of the paper to make the pouch blend in with the belly of the kangaroo.

Next, use construction paper to make a baby kangaroo. You can zip and unzip the pouch just like a mother kangaroo tightens and loosens the muscles of her pouch to protect her baby. You many even want to make various sizes of babies. When a joey first enters the pouch it is smaller than a bumblebee, has no hair, and its eyes are sealed shut. It grows and develops in the pouch. When it is big enough it leaves and reenters the pouch until it is too big to crawl back inside. Add your pouch and joey to your animal kingdom notebook.

What did we learn?

What is a marsupial? (An animal that gives birth to very tiny underdeveloped young. The young then spend the next several months developing in the mother's pouch.)

Name at least three marsupials? (Some of the more common include kangaroos, koalas, opossums, numbats, and Tasmanian devils)

How has God designed the kangaroo for jumping? (A kangaroo has large powerful hind legs, large hind feet, and long stretchy tendons that help conserve energy when hopping.)

Taking it further

About half of a kangaroo's body weight is from muscle. This is nearly twice as much as in most animals its size. How might this fact contribute to its ability to hop? (Large muscles are needed to provide the strength to hop long distances. So a kangaroo has very large leg muscles.)

How do you think a joey kangaroo keeps from falling out of its mother's pouch when she hops? (The nipple swells when the joey first attaches so it cannot slip off. Also, the pouch has muscles that can contract like a drawstring to keep the pouch closed.)

FUN FACTS

A red kangaroo can hop up to 27 feet in one leap.

A female kangaroo is ready to mate at about two years old. From that time on she will be nearly always pregnant. In fact she will often have a baby in her womb, a baby in her pouch and a youngster at her side all at the same time.

Kangaroos keep cool by panting like a dog. When it is really hot, a kangaroo will lick its forearms and the evaporation of the saliva will help to cool down the kangaroo.

Mammals Quiz

Lessons 1-7

Short answer:

1. What are the two main groups of animals? _____ _____

2. What are the five major groups of vertebrates? _____ _____
   _____ _____ _____

3. What are five common characteristics of mammals? _____ _____
   _____ _____ _____

4. What makes a marsupial different from other mammals? _____

5. What makes a vertebrate unique? _____

Answer True or False for each statement

6. _____ Animals can produce their own food.

7. _____ Dolphins are large fish.

8. _____ Marsupials give birth to tiny live babies.

9. _____ Baleen whales have large teeth.

10. _____ The elephant is the largest animal in the world.

11. _____ Monkeys have tails and apes do not.

12. _____ Marsupials live primarily in Australia and Tasmania.

13. _____ Some marsupials are meat eating animals.

14. _____ A lemur is a primate.

15. _____ Primates have eyes on the sides of their heads.

(Note: answers to all quizzes and tests are in Appendix A)

Lesson

8

Birds

Fine feathered friends

Supply list:

- 1 copy of "God Made Birds With Special Beaks" for each child (pg. 27)

Birds are some of the most interesting and easy to watch animals in God's creation. These warm-blooded feathered vertebrates can be found in every region of the world. There are approximately 9,000 different species of birds. Birds lay eggs and breathe with lungs. Most birds are excellent flyers, although some birds do not fly. God designed birds' bodies to be efficient flying machines. Birds have strong yet lightweight bones; many bones have hollow spaces to make them lighter. Birds also have a rigid or stiff backbone that supports the strong muscles used to move the wings.

With 9,000 different species of birds, it is helpful to group the birds by some common characteristics. Birds are often grouped as perching birds, birds of prey, water birds, game birds, tropical birds, and flightless birds. The design of the feet and beaks of these different groups reflects the different ecosystems or environments in which they live.

Approximately 60% of all birds are perching birds. Songbirds such as the thrush, robin, bluebird, and sparrow are just a few of the many perching birds. A few perching birds such as the hummingbird and woodpecker do not have songs. Perching birds have feet with 3 toes facing forward and 1 toe facing backward for grasping branches. Many have triangular-shaped pointed beaks for eating seeds and insects. Some of these birds, such as hummingbirds, have long narrow beaks for sucking nectar from flowers.

Birds of prey like eagles, hawks, falcons, and owls are designed to catch small animals such as rodents. They have very sharp eyesight, as well as hooked beaks and sharp talons (claw-like feet) that allow them to catch and kill their prey. Many birds of prey, particularly owls, also have very keen hearing that allows them to pinpoint prey in the dark.

Water birds, such as ducks, swans, and geese, are specially designed for life on and near the water. They have rounded beaks for catching fish and other food in the water and they have webbed feet for swimming. They also secrete oil that helps make their feathers water resistant.

Game birds are birds that are often hunted for meat. They have very strong flight muscles making them difficult to catch but good to eat once they are caught. These include wild turkeys, quail, and pheasant. Although ducks and geese are considered water birds, they are game birds as well.

Tropical birds include parrots, parakeets, and toucans. These birds live in the tropical rain forests. Most are very brightly colored and have large hooked beaks. They have similar feet to perching birds since they spend most of their time in the trees.

Finally, a few birds are flightless. These birds have wings but are not able to fly. Flightless birds include ostriches, emus and penguins. Most of these birds have strong legs and feet and can run or swim very swiftly.

Examining Beaks and Feet:

Examine the pictures in this lesson as well as pictures of birds in other books to get a better idea of how different birds' beaks and feet look. Then give your child a copy of the "God Made Birds With Special Beaks" worksheet and have him draw the different types of beaks and feet that God gave to birds.

Help your child recognize the various uses of the different shaped beaks and feet and how they help the birds to survive in their environment. Add this page to your animal kingdom notebook.

Other Optional Activities:

1. Put up a bird feeder and enjoy watching the birds come close on a regular basis.
2. Find abandoned nests and dissect them to find out what birds use to build their nests.

What did we learn?

How do birds differ from mammals? (They have feathers, eggs, wings, and can usually fly.)

How are birds the same as mammals? (They are both warm-blooded and breathe with lungs.)

How can you identify one bird from another? (By their size, shape, color of feathers, beak and feet design, calls, and songs)

Taking it further

What birds can you identify near your home? (Use a field guide to help you.)

Why might you see different birds near your home in the summer than in the winter? (Many birds migrate to live in a warmer area in the winter and a cooler area in the summer so different birds may be in your area at different times of the year.)

FUN FACTS

1. There are between 100 and 200 billion birds on the planet.
2. The largest bird is the ostrich, which can be up to 9ft. tall and weigh as much as 160 lb.
3. The heaviest flying bird is the Andean condor at 27 lb.
4. The smallest bird is the bee hummingbird from Cuba at 0.056 oz.

God Made Birds With Special Beaks:

For eating nectar   For eating fish   For eating prey

For eating bugs   For eating nuts and seeds

God Made Birds With Special Feet:

For swimming   For perching in trees

For catching prey   For walking or running

Charles Darwin
(1809-1882)

The name Charles Darwin can evoke strong emotions. Some people view him as one of the greatest scientists of the 19th century. Others see him as the man who destroyed our belief in God. Regardless of how you feel about evolutionism and creationism, it is important to know what Darwin did and to examine his findings in light of the Bible and in light of modern science. Charles Darwin was born in Shrewsbury, England in 1809. His father was a doctor and his mother was the daughter of the famous china maker Josiah Wedgwood.

After earning a degree in theology in 1831, he was selected to be part of a nature tour around the world. From 1831-1839, Darwin sailed from place to place on the HMS Beagle. At each place he visited, Darwin carefully studied and collected plants, animals, rocks, and fossils. He is most famous for his study of the finches of the Galapagos Islands. He discovered that different species of finches on each island had different beaks depending on the type of diet available there. This led him to seek an explanation for how such variety could occur. In 1859, after years of study, Darwin published his most famous work, *On the Origin of the Species*. In this work, Darwin suggested that changes within species were a result of natural selection – survival of the fittest. This idea was taken further to suggest that over time, these changes could result in a completely different species of animal. Later, in 1871, Darwin published a book called *The Descent of Man*, in which he suggested that man evolved from other species.

It is important to note that Darwin knew nothing about genetics or how traits were passed from one generation to the next. We know today that animals have a wide variety of traits that can be passed on through genes. We can see this variety in the many species of dogs that exist today. However, we also know that there is a limit to the amount of change that can occur genetically. Evolutionists claim that mutations in the genes produce changes that would not normally occur in an animal, and that if these mutations are beneficial, they are passed on to the offspring. However, no one has ever observed a beneficial mutation. All observed mutations have been determined to be harmful or neutral to the animal.

We can observe adaptation, such as Darwin observed with his finches, but we do not see changes that change one species into another. Furthermore, no fossil evidence has been found to show a progression of changes from one species into another. What we observe in both nature and in fossils is that a finch is still a finch and a man is still a man. God created each type with a wonderful capacity for variety, but there is no evolution from one species to another.

Flight

How do those birds do that?

Supply list:

> - 1 copy of the "God Designed Birds To Fly" for each child (pg. 32)
>
> - 1 or more bird feathers
>
> - Magnifying glass

Although some birds are flightless, most birds are designed for flight. To watch a bird soaring into the sky is a marvelous thing. For centuries man sought to imitate birds, but only in the last 100 years or so has man truly begun to understand how perfectly designed the bird's body is for flight.

Birds have very strong breast muscles attached to a wide sternum, and they have stiff backbones to withstand the forces of flight. The muscles help move the wings smoothly and efficiently. Birds also have a special respiratory system that allows them to breathe air through lungs into a system of air sacs that extract much more oxygen than any other animal's respiratory system can. Birds have hollow spaces in their bones, making them extremely light for their size. These and many other special features aid birds in flying.

By far the most important feature to help birds fly is the design of their wings and feathers. Their wings are shaped like an airfoil, forcing air to flow more quickly over the wings than under them, thus creating lift.

Birds have three kinds of feathers, each with a different function. Soft fuzzy down feathers provide insulation near the bird's body. Over these are the contour feathers that cover the bird's body. All contour feathers point toward the tail, making air flow smoothly over the bird's body. Flight feathers give the wing the needed shape for flying. The feathers are designed with a hook and barb system to help the feather maintain its shape. If a feather gets pulled apart the bird can zip it back up with its beak

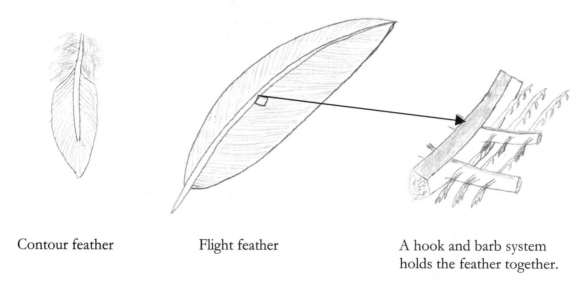

Contour feather   Flight feather   A hook and barb system
                                                          holds the feather together.

A bird's wing has 3 sets of flight feathers. The primary feathers are attached near the end of the wing. The secondary feathers are attached in the center, and the tertiary feathers are attached to the upper wing, close to the body. Movement of the flight feathers makes tiny changes in the shape of the wing to compensate for changing air conditions. Contour feathers on the front of the wing make a smooth surface over which the air can easily flow.

Wing covered with feathers   Wing showing bone structure

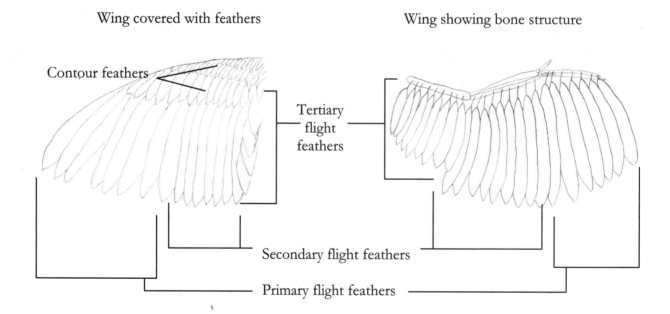

Contour feathers

Tertiary flight feathers

Secondary flight feathers

Primary flight feathers

Finally, the bird's tail serves as a rudder. By moving the tail from side to side, the bird can steer or change direction in the air. God designed every part of the bird for efficient flight.

Examining a Bird's Feather:

Examine a bird's feather with a magnifying glass. Notice the barbs that hold the feather together. A bird can "zip up" its feathers with its beak if they get pulled open. This is called preening. Birds spend some time every day fixing or preening their feathers.

Wing and Feather Worksheet:

Give your child a copy of the "God Designed Birds To Fly" worksheet and have him draw an airfoil, label the feathers on the bird's wing, and draw an example of the structure of a feather. Glue the bird's feather to the sheet. Add this page to your animal notebook.

What did we learn?

What are some ways birds are designed for flight? (They have strong breast muscles, rigid backbones, hollow bones, efficient respiratory systems, feathers, and wings)

What are the 3 kinds of bird feathers? (Down, contour, and flight feathers)

How does a bird repair a feather that is pulled open? (By preening – running the feather through its beak to re-hook the barbs)

How does a bird's tail work like a rudder? (It is moved from side to side to help steer.)

Taking it further

Why can't man fly by strapping wings to his arms? (Man is not designed for flight. He does not have the strong breast muscles and stiff backbone needed. Humans are also too heavy to lift themselves with their arms.)

How do birds use their feathers to stay warm? (Birds can fluff up their feathers and trap air under and between them. The heat from their bodies warms the trapped air, creating a barrier between their bodies and the cold air around them.)

How is an airplane wing like a bird's wing? (They both have the same airfoil shape that allows the air flowing over the wing to create lift. Airplane wings are also designed with the ability to change shape to adapt to different conditions, just like birds' wings.)

God Designed Birds To Fly.

Their wings are designed like airfoils to achieve lift:

They have 3 sets of flight feathers on their wings in order to change the shape of the wing during flight: (primary, secondary, and tertiary)

Feathers are designed to be strong, light, and flexible. They have a hook and barb design to help them keep their shape.

Lesson 10

The Bird's Digestive System

They sure eat a lot

Supply list:

- 1 copy of "God Designed the Bird's Digestive System" for each child (pg. 36)

- Optional: Owl pellet

Have you ever noticed that when birds are not flying they are almost always eating? This is because flying requires a lot of energy. In addition, warm-blooded animals need a lot of food to keep their bodies at a constant temperature. God has created birds with a special digestive system to help them be able to fly and regulate their temperature.

First, a bird's digestive system works very quickly. A bird can digest its food in as little as 30 minutes to 3 hours. The human body takes several hours to as much as two days to completely digest its food. Second, a bird's digestive system is very efficient at extracting the nutrients it needs.

Birds do not have teeth, so God designed them with a digestive system that digests food that has not been chewed. A bird swallows its food, which enters its esophagus. The food is then held in a sac called a crop to be released into the stomach at a constant rate. This allows a bird to eat quickly then fly to a safe place for digestion. The food then goes into a small stomach where digestive juices are added. The food then enters the gizzard. This organ is very rough inside and grinds up the food. The gizzard compensates for the bird's lack of teeth. After that, the food enters a small intestine where the nutrients are extracted after which, the waste goes through the cloaca to be eliminated.

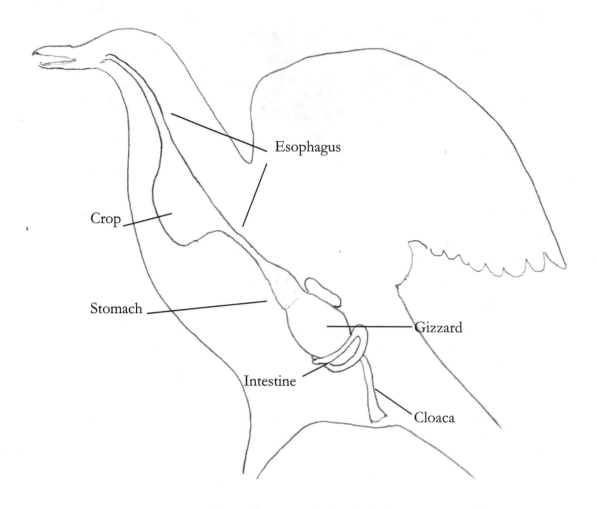

Bird's Digestive System

This special digestive system helps birds have the necessary energy for flight.

Bird's Digestive System Worksheet:

Have your child draw in and label the parts of the bird's digestive system on the "God Designed the Bird's Digestive System" worksheet. This can be added to the animal notebook.

Optional Activity - Dissect An Owl Pellet:

Owl pellets can be ordered from many science supply stores. (See Appendix B for possible suppliers.) An owl swallows its prey whole and later spits up a pellet of indigestible fur and bones. You can obtain pellets that have been sterilized and dissect them to see what the owl had for dinner by matching the bones to a chart that can also be purchased. This is a fascinating project.

What did we learn?

How is a bird's digestive system different from a human digestive system? (A bird has a crop and a gizzard; humans do not. A bird does not have a large intestine. A bird's digestive system digests food much more quickly.)

How does a bird "chew" its food without teeth? (God designed birds with an organ called a gizzard, which grinds the food up internally. In addition, some birds swallow small stones or pebbles that help to grind up the food as well.)

What purposes does the crop serve? (It holds the food so a bird can eat quickly. It then releases the food to be digested in a constant stream to provide a more constant source of energy.)

Taking it further

How does a bird's digestive system help it to be a better flyer? (Because the food is digested quickly and efficiently, more energy is available for flying. Because the food is digested at a constant rate, a steady source of energy is provided for extended flying.)

FUN FACT

In addition to having a special digestive system, a bird also has a special respiratory system that allows it to fly for long distances. A bird's respiratory system is extremely efficient in extracting oxygen from the air, since it is designed in such a way that the air actually passes through the lungs twice before being exhaled. Also, a bird's respiratory system helps cool the bird during flight. The cool fresh air not only flows through the lungs and into air sacs, but also flows through tubes from the air sacs to hollow spaces in the bones. Heat is transferred to the air and leaves the body when the air is exhaled. This helps keep an active bird from overheating.

God Designed the Bird's Digestive System

Birds do not have teeth so God designed them with a special digestive system.

Lesson

11

Amphibians

Air or water?

Supply list:

- Your imagination

As we have learned, there are five groups of vertebrates (animals with backbones). The first two groups, mammals and birds, are warm-blooded. Their body temperatures remain constant. The others, amphibians, reptiles, and fish, are cold-blooded. This means their body temperatures go up when the temperature around them is warm and down when their environment gets colder.

Amphibians are cold-blooded animals that have smooth moist skin. They generally live in very moist areas or near the water to keep their skin from drying out. They lay eggs. But the unique thing about amphibians is that they spend the first part of their lives in the water using gills to breathe. Then their bodies change and they develop lungs that allow them to breathe air. The word amphibian means "on both sides of life," reflecting this change (or metamorphosis).

The three major groups of amphibians are: frogs and toads, salamanders, and caecilians. The vast majority of amphibians are frogs and toads. Frogs and toads come in many sizes and colors. Frogs usually have smooth moist skin and toads usually have more dry bumpy skin. Aquatic frogs live in or near the water. Tree frogs live in tropical forests in the trees. Tree frogs are usually much more colorful and often are poisonous. Tree frogs also have

suction cups on the bottoms of their feet to allow them to climb trees very quickly. Toads often live further from the water but most must return to the water to reproduce.

Salamanders have long thin bodies and tails. They might be confused with lizards, but their skin is smooth and moist, unlike a lizard's skin which is dry and scaly. Also, as babies, salamanders look very different from lizards. They spend the first part of their lives in the water in a larval stage before they change into the more familiar adult form.

The smallest group of amphibians is the caecilians. Caecilians are legless amphibians that resemble worms. They are long and thin and usually burrow in the ground. Caecilians live in the tropical forests of South America, Africa, and Southeast Asia. Because they spend most of their lives under ground, people seldom see them.

Warm-blooded/ Cold-blooded:

Have one child pretend to be a warm-blooded animal such as a dog and have another child pretend to be a cold-blooded animal such as a frog. Let each child act out what that animal might be doing if the temperature around it was 15 degrees Fahrenheit. Remind your children that cold-blooded animals cannot be active in cold weather and they often go into hibernation.

Now have your children act out what their animals might do if the temperature was 65 degrees Fahrenheit, and again if the temperature was 95 degrees Fahrenheit. Remind your children that warm-blooded animals must find ways to keep their bodies warm in the cold weather, and cool in the hot weather. Remind them that cold-blooded animals must find shade when the weather is too hot.

What did we learn?

What are the characteristics that make amphibians unique? (They spend part of their lives in water breathing with gills and part of their lives on land breathing with lungs. They are also cold-blooded, have smooth moist skin, and lay eggs.)

How can you tell a frog from a toad? (In general, frogs have smooth moist skin, while toads have dry bumpy skin.)

How can you tell a salamander from a lizard? (Salamanders have smooth skin and lizards have dry scales on their skin. Also, salamanders go through metamorphosis and lizards do not.)

What is the difference between warm-blooded and cold-blooded animals? (Warm-blooded animals regulate their body temperature – it stays the same regardless of the surrounding temperature. Cold-blooded animals cannot regulate their body temperature – it goes up and down with the surrounding temperature.)

Taking it further

What advantages do cold-blooded animals have over warm-blooded animals? (They don't have to eat as often, and can usually survive a broader range of temperatures.)

What advantages do warm-blooded animals have over cold-blooded animals? (Cold-blooded animals' activities are more restricted by temperature extremes. A warm-blooded animal can still be quite active in very cold or very warm weather.)

Why are most people unfamiliar with caecilians? (They spend most of their time underground and live only in tropical rain forests, so most people never see them.)

Lesson 12

Amphibian Metamorphosis

Making a change

Supply list:

- 1 copy of "Amphibian Lifecycle" worksheet per child (pg. 42)

- Book about frogs -- showing metamorphosis would be helpful

- Optional: tadpoles, tank, food for raising a frog

Amphibians do what no other animals do. They start out in life getting oxygen from water using gills. Then they slowly change into an adult with lungs to get oxygen from the air. The most familiar amphibian is the frog. Although all amphibians go through this metamorphosis or change, we will examine the changes a frog experiences to better understand this process.

Most frogs go to the water to reproduce, even if they do not live in or near the water the rest of the time. Their eggs are laid in a mass in the water. A jelly-like substance coats and protects the eggs until they hatch. Eggs usually hatch in 6-9 days.

What hatches from the egg is called a tadpole (or sometimes a pollywog). The tadpole is the larva stage in the frog's lifecycle. This "infant" frog has a tail for swimming and looks a little like a small fish (see the picture above). It lives in the water and spends most of its time eating and growing. The tadpole has gills, which are organs that transfer oxygen from the water into the animal's blood stream.

In a matter of weeks the tadpole begins to change noticeably. This change is called metamorphosis. During metamorphosis hind legs begin to grow at the base of the tail. The front legs begin to form. Eventually the tail shrinks away. At the same time the legs are forming, the frog begins to develop lungs. Lungs are organs that transfer oxygen from the air to the animal's bloodstream. Until the lungs are fully developed, the frog uses gills to extract oxygen from the water. Once the lungs are ready the gills begin to disappear and the tadpole has transformed into a frog! Most adult frogs leave the water and spend the majority of their adult lives on land and return to the water only to lay eggs. A few types of frogs continue to live close to the water throughout their lives. This transformation from water dweller with gills to land dweller with lungs is what makes amphibians unique.

Amphibian Lifecycle Worksheet:

After looking at pictures of a frog changing from a tadpole into an adult, have your child draw a picture representing each stage of an amphibian's lifecycle on the "Amphibian Lifecycle" worksheet. Add this sheet to your animal kingdom notebook.

Optional Activity – Grow a Frog:

The best way to appreciate the metamorphosis of a frog is to get a tadpole and watch it change day by day. Science supply catalogs often sell Grow-a-frog kits that come with a live tadpole and pet stores sometimes sell tadpoles. (See Appendix B for possible suppliers.) Or if you live near a pond you might be able to catch some tadpoles to raise and then release the grown frogs.

What did we learn?

Describe the stages an amphibian goes through in its lifecycle. (It begins as an egg, and then it hatches into a larva. In a frog, this is the tadpole stage. Then, it slowly changes into an adult. This is the metamorphosis stage in which lungs develop and gills disappear, and the creature changes its shape from a water dweller without legs to a land dweller with legs.)

What are gills? (They are special organs on the sides of water animals that extract oxygen from the water as water passes over or through them.)

What are lungs? (They are special organs that extract oxygen from the air as air passes through them.)

Taking it further

Does the amphibian lifecycle represent evolution? (NO! Evolution says that one species changes into another. A frog is still a frog even when it is a tadpole. A tadpole always changes into a frog. It does not grow up to be a bird or a mammal or even a salamander. It is always what God made it to be, even if its infant form is significantly different from its adult form.)

Amphibian Lifecycle

Egg

Larva

Metamorphosis

Adult

Reptiles

Scaly animals

Supply list:

- Paper

- Pictures of reptiles

- Sequins or flat beads

- Glue

Ever since the serpent tempted Eve in the garden of Eden, reptiles have held a strange fascination and often fear for humans. Reptiles are vertebrates with dry scaly skin. They lay eggs, breathe with lungs, and are cold-blooded. Most reptiles have nictitating membranes, which are clear eyelids that cover and protect the eyes. These membranes are needed so the animal can still see even if it is in harsh conditions such as a wind storm in the desert or underwater.

The four major groups of reptiles are: snakes, lizards, turtles, and crocodiles. In addition, there is one species of tuatara. Tuataras are similar to lizards but have a slower metabolism. Scientists at one time believed these creatures were extinct. They now call them living fossils because what they thought was a creature that could only be found in fossils is now a living creature. This puts a hole in many evolutionary ideas because many scientists believed the tuataras no longer existed because they had evolved into something else. But since scientists have found living specimens, the tuataras obviously did not evolve into a different creature.

Of the 6000 different kinds of reptiles, about half are lizards. Another 2700 species are snakes. There are only about 240 kinds of turtles and only 21 kinds of crocodiles and alligators.

Reptiles live in all parts of the world. Because they are cold-blooded, many reptiles in hot tropical climates are nocturnal. This means they sleep during the day and are active at night. Reptiles can also live in very cold climates. But because their bodies slow down so much in the cold, they go into a sort of hibernation until the weather warms up.

Scaly Picture:

Have your child draw an outline of a reptile. Then have him glue sequins wherever the creature would have scales on its body. Turtles only have scales on their legs, feet, tail, neck and head. Alligators have very large scales on their bodies. Snakes have scales over their entire bodies. Include this picture in your animal kingdom notebook.

What did we learn?

What makes reptiles different from amphibians? (Reptiles have scales and amphibians do not. Also, reptiles have lungs all their lives and do not go through metamorphosis.)

How do reptiles keep from overheating? (They stay in the shade or other cooler places during the hottest part of the day. Many sleep during the day and are only active at night.)

Taking it further

What would a reptile do if you dug it out of its winter hibernation spot? (It would appear dead. It would not move or eat. If you brought it inside and it warmed up, then it would seem to come alive, though it is actually alive even in its hibernating state.)

Snakes

Those hissing slithering creatures

Supply list:

- An open area on the floor for moving about

The vast majority of reptiles are snakes and lizards. Snakes are the only legless reptiles. Snakes all have the same general shape, long and round, but can be as short as 5 inches or as long as 30 feet.

Snakes have eyes on the sides of their heads but no eyelids. Instead, they have clear scales that fit over their eyes. They do not have external ears. Instead, they sense vibrations and low noises through their lower jaw, which sends the vibration to an inner ear. Snakes have nostrils that help with smell but, in addition, each time they flick their tongues, they pick up scent particles. These particles are then touched to the Jacobsons' organ in the top of the mouth. This organ is very sensitive to smell and allows the snake to follow the scent of its prey.

Snakes do not need to eat very frequently since they are cold-blooded. A warm-blooded animal eats frequently because it requires more energy to maintain its body temperature. Cold-blooded animals do not maintain a specific temperature so they do not need as much energy as a warm-blooded animal. Therefore, they do not have to eat as often.

Snakes eat many different kinds of prey, from insects to large mammals, depending on the size of the snake. Snakes cannot chew or tear their food; instead, they swallow it whole. Snakes are able to do this because their lower jaws are not permanently attached to their skulls. They can disconnect their lower jaws and stretch their mouths around something much larger than the diameter of their own bodies. After swallowing its meal, the snake's strong muscles squeeze its prey as it begins digestion.

Most snakes fall into one of three groups: constrictors, colubrids, and venomous snakes. Constrictors are found mostly in tropical jungles. Constrictors overcome their prey by wrapping around and squeezing them. As the victim exhales, the constrictor squeezes tightly. After a few minutes, the prey has suffocated, and the snake then eats it. Pythons and boas are some of the most familiar constrictors.

Over 2/3 of all snakes are colubrids. Colubrids are found in most parts of the world. Most of these snakes are non-poisonous and many are useful for keeping down the rodent population. Bull snakes, rat snakes, and garter snakes are some common colubrids.

The most feared snakes are the venomous or poisonous snakes. These snakes have fangs that are used to inject venom into their prey. The venom attacks the nervous system, circulatory system, or both. The prey is usually paralyzed and stops breathing in a matter of minutes. Then the snake can eat its meal without a struggle. Many, but not all, venomous snakes are poisonous to humans. Some well-known venomous snakes include cobras, rattlesnakes, and coral snakes.

Slithering Like a Snake:

Have your child lie on the floor and try to move in the following ways. It may not be easy to match the movements of a snake but your child will have fun trying.

Snakes move in one of four ways.

1. Lateral Undulation – Sideways waves
The most common way for a snake to move is in this S-shaped squiggling. The snake's body moves in curves from side to side. Snakes can move on land or in the water this way. Have your child lie on his stomach, then try squiggling from side to side without using his arms.

2. Rectilinear – Straight line
Snakes that move rectilinearly stretch then contract their bodies in a straight line. They use their scales to help anchor one part of their bodies while they move another part. Have your child lie on his stomach, pull his knees up to his chest, and then stretch back out.

3. Concertina
This is a slinky type of movement where the snake coils up then uncoils like a spring. Have your child lie on his side, bend at the waist and knees, then push with his feet to spring forward.

4. Side Winding
A sidewinder anchors its head and tail, moves its body sideways, then moves its head and tail to match the body. The result is a diagonal movement to the direction the snake is facing. Have your child lie on his stomach, move his hips sideways, move his head and shoulders to line up with his hips, and then repeat this motion.

What did we learn?

How are snakes different from other reptiles? (They have no legs.)

What are the three groups of snakes? (Constrictors, colubrids, and venomous snakes)

How is a snake's sense of smell different from that of most other animals? (It uses its tongue to collect scent particles, and then touches them to an organ called the Jacobsen's organ inside its mouth.)

What is unique about how a snake eats? (It swallows its food whole and can eat something larger than its body diameter by unhooking its jaw and stretching its mouth very wide.)

Taking it further

How are small snakes different from worms? (Snakes have backbones, scales, and well-developed eyes. Worms do not have any of these. Also, snakes have much more complicated internal systems.)

If you see a snake in your yard, how do you know if it is dangerous? (You should learn to identify snakes using a guidebook or other resource. Unless you have your parent's permission, you should never approach a snake.)

Rattlesnakes

It's a sound that puts fear in every heart: the gentle rattle coming from the grass. Rattlesnakes can be a frightening and dangerous sight. Yet, they are very interesting creatures. Rattlesnakes are the largest snakes that live in the United States. They can be found in nearly every part of North and South America.

Rattlesnakes are different from all other snakes because they have a rattle on the end of their tail. Each time a rattlesnake outgrows and sheds its old skin, its rattle gains another ring. These rings hitting each other as the snake shakes its tail are what give the rattlesnake its distinctive sound.

Rattlesnakes are vipers and are therefore poisonous. They kill their prey by injecting it with venom. It bites its prey very quickly then pulls back. The animal will run away but will soon die. The poison paralyzes the animal and then begins to break down its body before the snake even swallows it. The snake then uses its sense of smell to track down the animal and swallows it whole shortly after it dies.

Rattlesnakes seldom bite people. They usually give a warning by shaking their tails if someone approaches and frightens them. If you should come upon a rattler be sure to stay at least 10 feet away. A snake can only strike about half the length of its body, so if you stay back it cannot bite you. A rattler will usually not bite someone unless that person persists in getting too close. Rattlers bite about one thousand people each year in the United States. Nearly everyone who is bitten survives if they go to a hospital for treatment. Doctors have developed a serum called antivenin, which is made from the venom of rattlesnakes and helps by breaking down the toxins in the venom.

Although rattlesnakes can be harmful to people, they can also be very helpful. Snakes are some of the best mice and rat hunters and are helpful in keeping down the pest population.

- There are 70 different kinds of rattlesnakes
- A rattlesnake's rattle is made from keratin – the same material your fingernails are made of.
- Rattlesnakes hatch their young inside their bodies before giving birth.
- A rattlesnake can eat an animal as big as a five pound rabbit.
- Rattlesnakes have no eyelids. In fact no snakes have eyelids.
- The smallest adult rattlesnake is the pygmy rattlesnake, which grows to only 18 inches long.
- The largest rattlesnake is the Eastern diamondback, which grows up to 8 feet long.
- The most poisonous rattlesnake in the United States is the Western diamondback.
- If a rattlesnake loses a fang it can grow a new one.
- About 10 people die each year in the United States from rattlesnake bites.

Lesson

15

Lizards

Chameleons and Gila monsters

Supply list:

- Paper

- Markers, colored pencils, or paints

- Face paint

L izards are the largest group of reptiles. They have long thin bodies with legs that attach to the sides of their bodies. They have tapered tails. Their feet have claws. And they are covered with scales.

Lizards can be found in nearly every climate and ecosystem. They range in size from a few inches to 12 feet. Most lizards are not dangerous to humans, but the Gila monster is venomous and Komodo dragons can cause infection if they bite a human.

Lizards have various ways to protect themselves from predators. The chameleon can change colors to match its surroundings. Some chameleons can change only their color, while others can change both their color and pattern. Lizards that cannot change color have other forms of self-defense. The horned lizard has sharp spikes on its head and back to protect it from its predators. The chuckwalla crawls into a crack in the rocks when it feels threatened, then fills its body with air, making it nearly impossible to get it out of the crack. Finally, some lizards have the ability to shed or break off their tails if a predator grabs on. Later a new tail will grow in its place. The predator gets the tail but the lizard gets away.

A few lizards eat plants, but the majority of lizards eat insects. This makes them a welcome visitor in many homes, especially in the tropical areas. Komodo dragons, some of the largest living reptiles, eat dead animals. They have a keen sense of smell and will cross large distances to reach a decaying carcass.

Animal Camouflage:

Have your child draw or paint a picture showing how a chameleon can blend into its surroundings. For example, show a chameleon on a rock. Make the chameleon have a similar color and pattern to the rock, or show it sitting in a tree with similar colors to the leaves. If you have sequins left from the reptiles project your child could glue them on the chameleon if they are similar colors to the surroundings. **Add this page to your animal kingdom notebook.**

People Camouflage:

Discuss how soldiers use camouflage paint to cover their skin and camouflage clothing to help them blend in with their surroundings. Use face paint to cover your child's skin to help them blend in with the trees and bushes or other environment you may choose. If you like, you can take a picture of your camouflaged child and include it in the animal notebook.

What did we learn?

Horny lizards are short compared to many other lizards and are often called horny toads. What distinguishes a lizard from a toad? (Toads do not have scales but lizards do. Also, lizards do not have gills when they are young, nor do they experience metamorphosis, but toads do.)

List three ways a lizard might protect itself from a predator. (It could change its color, crawl into a crack in a rock and inflate its body, or break off its tail to escape.)

Taking it further

Why might some people like having lizards around? (They eat insects and do not harm people.)

How does changing color protect a lizard? (It makes it hard for the predator to see it.)

What other reasons might cause a lizard to change colors? (To attract a mate or scare off competitors)

Turtles and Crocodiles

Turtle or tortoise, crocodile or alligator--how do you tell?

Supply list:

- 1 copy of "How Can You Tell Them Apart?" for each child (pg. 54)

- Tape – cloth or first aid tape would probably work the best

- Sink

Turtles are the only reptiles with shells. Unlike many sea creatures such as crabs, which can be removed from their shells, turtles' shells are an integrated part of their bodies, not just a home they live in. The shell provides protection for the turtle's internal organs. Also, many turtles can pull their heads, tails, and legs inside their shells when they feel threatened.

Like all reptiles, turtles have scales covering their skin, are cold-blooded, and lay eggs. Turtles can be found in most warm climates. In some areas with cold winters, turtles will burrow underground and hibernate until warmer weather arrives.

The term "turtle" generally refers to the turtles that live in or near water. Fresh water and sea turtles are equipped with webbed feet or flippers for swimming. Turtles that live mainly on land are generally called tortoises. Tortoises have claws instead of webbed feet, as well as  short sturdy legs designed for walking on land. Both turtles and tortoises lay their eggs on land.

The group of reptiles with the smallest number of species is the crocodiles. This group includes crocodiles and alligators. These can be the largest reptiles with bodies up to 25 feet long. It is often difficult to distinguish a crocodile from an alligator. Alligators have wider snouts and all of their teeth are covered when their mouths are closed. Crocodiles have longer, narrower snouts and some of their teeth, usually one on the bottom on each side, stick out even when their mouths are closed.

Crocodiles and alligators are generally found in warm tropical climates. They live near water and have webbed feet and strong tails that enable them to be good swimmers. The eyes and nostrils are located on the top of the head and snout, allowing the animal to float with most of its body underwater while it waits for prey to come close enough to attack. A floating crocodile resembles a fallen log floating in the water. When the prey is close, the crocodile snaps its jaws around the victim and drags it under water, where it holds the prey until it drowns. Crocodiles generally eat turtles, fish, waterfowl and other small animals. But some larger crocodiles will attack larger animals as well.

Like most reptiles, crocodiles lay their eggs on land. But unlike other reptiles, when the babies are about to hatch, the mother carries the eggs to the edge of the water in her mouth. After they hatch, the mother protects her young for several weeks. Adult crocodiles will eat young crocodiles, so after they leave their mother's protection, they stay away from adults until they are fully grown.

Turtle Feet – Tortoise Feet:

To better understand the difference in turtle and tortoise feet, try this experiment. Tape all the fingers on one of your child's hands together to represent a turtle's flipper. Have your child make a claw with their other hand. Fill up the kitchen sink with water or try this in a bathtub or swimming pool. Have your child try pushing the water with his flipper and his claw. Which hand was able to push the most water? The flipper is better designed for swimming. Since the turtle spends most of its time in the water, the flipper is a better design than a claw-like foot. God designed each species to best survive in its environment.

How Can You Tell Them Apart Worksheet:

Have your child complete the "How Can You Tell Them Apart?" worksheet by drawing pictures showing how to tell turtles from tortoises and crocodiles from alligators. This worksheet can be included in the animal kingdom notebook.

What did we learn?

Where do turtles live? (In the water)

Where do tortoises live? (On land)

How does the mother crocodile carry her eggs to the water? (In her mouth)

Why can't you take a turtle out of its shell? (Its shell is part of its body)

How do crocodiles stalk their prey? (They float in the water, wait for prey to approach, and then clamp their jaws around the prey and drag it under the water to drown it before eating it.)

Taking it further

Why might it be difficult to see a crocodile? (When it is floating in the water, it looks very much like a fallen log. This is how it tricks its prey into coming close enough to be eaten.)

FUN FACT

Have you ever heard of crying crocodile tears? Crocodiles often secrete a liquid from their eyes as they are wrestling with their prey. This gives the appearance that they are crying for their victims. Thus, "crocodile tears" refers to insincere sorrow.

FUN FACT

No one knows for sure how long turtles and tortoises live. The Galapagos giant tortoise is believed to live for over 150 years, and many other turtles and tortoises live to be well over 50 years old.

How Can You Tell Them Apart?

How can you tell if you are looking at a turtle or a tortoise?

1. Look at its habitat. Does it live in the water or on the land? Turtles spend most of their time in the water and tortoises live on the land.
2. Look at its feet and legs. Turtles have webbed feet or paddle-like legs. Tortoises have short stocky legs with clawed feet.

Turtle Feet   Tortoise Feet

How can you tell if you are looking at a crocodile or an alligator?

1. Look at the shape of its snout. Crocodiles have long thin snouts. Alligators have shorter, more rounded snouts.
2. Look at its mouth from the side when it is closed. Can you see any teeth sticking out? If you cannot see any teeth you are looking at an alligator. If you can see one or more teeth sticking out on the bottom you are looking at a crocodile.

Top view of snouts

Crocodile   Alligator

Side view of snouts

Crocodile   Alligator

When Did the Dinosaurs Live?

When did the dinosaurs live? If you listen to many "experts" they will tell you that all dinosaurs lived and died out about 65 million years before humans were around. But if you read the Bible, Genesis chapter 1 says God made all the animals that live in the water, and all that fly, on the fifth day. On the sixth day God made all the animals that live on land, including dinosaurs and people. The Bible, God's Word, says that dinosaurs and man were created on the same day and lived together. So which is it?

The only way to scientifically prove something is to either observe it or recreate it. When it comes to the beginning of life, we can neither observe nor recreate it. Since there is no proof by scientific standards, we must look at the evidence.

The Bible tells us that dinosaurs were created on the sixth day along with man. But, since we don't see dinosaurs much these days, we may be convinced that they never lived with man. However, there are other written records of dinosaurs living with man. Job 40:15-24 talks about the behemoth with a tail that sways like a cedar and bones that are like tubes of bronze. It talks about his strength and the power in the muscles of his belly. This doesn't describe any animal we have living today but it does sound a lot like the brachiosaurus. This was a description of an animal that Job was familiar with, thus showing that man lived at the same time as this animal.

The Bible is not the only source of evidence for man and dinosaurs living together. There are many interesting examples that have been documented in the last hundred years that would indicate that dinosaurs not only lived with man at one time, but that they might still be living today. Let's look at a few of these cases.

In 1915, a German U-boat sank a British steamer. Shortly after the steamer went under the water it exploded. Within a few minutes parts of the wreckage, along with a sea animal, came to the surface. The captain, chief engineer, navigator, and helmsman from the U-boat all witnessed the sea animal. The sea animal was about 60 feet long, with a crocodile shape. It had four limbs with powerful webbed feet and a long tapering tail. This sounds very much like a sea dinosaur still existed in 1915.

In 1977, a Japanese fishing boat brought a smelly "monster" up in its nets from the depths of 900 feet. The fishermen took pictures of it, measured it (32 feet long and weighed over 4000 lbs), and took tissue samples. After this, they promptly pushed it back into the water before it could contaminate their "real catch". When the Japanese scientists examined the data they said it was a Plesiosaur that appeared to have died only a few weeks prior to being caught in the net.

In 1993, evolutionist scientists at the University of Montana found Tyrannosaurus rex bones that were not totally fossilized. These bones still had DNA in their centers. This DNA may have been from the T. rex's blood. If the bones were 65 million years old, the whole bone would have been fossilized, including the blood. This evidence indicates that the dinosaur may have been living in recent times.

In Alaska, several unfossilized dinosaurs bone sites have been found. Some evolutionists say this is because the bones have been under ice for millions of years. However, these same scientists claim that Alaska was much warmer millions of years ago, thus allowing the dinosaurs to live there. Both situations cannot be true. It is much more likely that the bones are from dinosaurs that have not been dead very long.

In the northern region of the Congo, a place where few outsiders have ever gone, there is said to live an animal much larger than an elephant. The pygmies call this animal Mokele Mbembe (mo-KEE-le MEM-be) and say it is brownish-gray with smooth skin. It has a very large and powerful tail. When the pygmies were shown pictures of different animals they pointed to the apatosaurus-looking one. If these dinosaurs died out so long ago, why would these people identify it as the animal living in their jungle?

Scientists have found dinosaur footprints preserved in stone. One footprint that evolutionists have tried to ignore is one that has a man's footprint imbedded in the middle of the dinosaur's footprint. The fossilized footprints had to have been made when the ground was wet and was then covered over quickly before the footprints could be washed away. This shows that the man and the dinosaur had to have passed through the same area within a very short time of each other. All of these examples provide very strong evidence that dinosaurs lived at the same time as man and that some dinosaurs may still be living today.

Another piece of evidence indicating that dinosaurs existed with man is the fact that almost every culture has stories of "dragons" (the European name). These great animals are described as giant lizards that destroyed small towns. Brave knights often had to fight them to save the towns. When you read the descriptions of these animals, they sound a lot like dinosaurs (which means terrible lizard). It is very unlikely that so many different cultures would have the same stories in their history if there wasn't some truth in them.

Now you may want to know what happened to the dinosaurs. If they were so big and powerful why aren't they still plentiful on the earth? This is one place Creationists and Evolutionists agree, something happened to change the Earth's climate. Some evolutionists say a meteor hit the Earth spilling millions of tons of dirt into the air causing an ice age. Others say that a meteor may have hit the Earth causing a flood. In Genesis chapters 6-8, the Bible says that God destroyed the Earth with a flood because of man's wickedness. The Earth was changed forever and the dinosaurs couldn't adapt very well.

A world-wide flood shows up in more then 270 different legends around the world. In Hawaii there is a legend that long after the first man died the world become very evil and there was only one good man left. His name was Nu-u. He made a very large canoe with a house on it, and filled the canoe with animals. Then the waters covered the Earth and killed the people. Only Nu-u and his family survived.

In China they have writings about a great flood that covered the whole Earth and killed everything except one man and his three sons and three daughters. They even have an ancient temple with a painting of the boat in the raging waters. The painting shows a dove with an olive branch in his beak flying toward it.

The Toltec Indians in Mexico have stories that the first world lasted 1716 years and was destroyed by a great flood. They say only a few men escaped. After the flood the men multiplied and started building a great tower to give them a safe place in case it flooded again. But their language become confused, so the different groups that could understand each other wandered off to different parts of the world.

There is very strong evidence that a world-wide flood happened. A great flood like this would change the climate and cause most of the animals to die and many to be covered with mud before the bones had time to decay. This would result in a multitude of fossils. A world-wide flood would also set up the right conditions for an ice age. This is exactly what we see in the fossil evidence.

However, since the ark had every kind of animal in it, all the kinds of animals including dinosaurs were saved. The dinosaurs may have been babies at the time of the flood, but they survived, and some of them lived on for a long time. The ice age and other climatic changes caused most of the dinosaurs and many other types of animals to die out.

By examining the evidence, we see that it supports the biblical explanation much more accurately than the evolutionistic explanations. Which will you believe?

Fish

Do fish go to school?

Supply list:

- Goldfish snack crackers
- Paper
- Glue
- Colored pencils

One of Americans' favorite pastimes is fishing. Fish can be found in the smallest ponds and streams, in large lakes, and throughout the oceans. There are over 22,000 species of fish, making them the most diverse group of vertebrates. Fish are vertebrates that live in the water. They have scales, and they breathe oxygen from the water using gills. Fish are cold-blooded, allowing them to survive even in very cold climates. Most fish reproduce by laying eggs, although a few give live birth. Most fish live in either just salt water or just fresh water but a few can survive in both salt and fresh water environments.

Most fish are bony fish, meaning they have bony skeletons. Bony fish have fins and tails that make them excellent swimmers. When you think of a fish you probably think of a fish such as a trout, bass, or goldfish. But about 5 % of all fish do not have rigid bones. Instead they have flexible skeletons made from cartilage. Cartilaginous fish include sharks, rays, and lampreys.

Even though dolphins and whales appear to be a lot like fish, there are a few very important differences. One of the most distinct differences is in how they breathe. Dolphins, whales, and other aquatic mammals must periodically go to the surface for fresh air, which they take into their lungs. Fish, on the other hand, get their oxygen from the water. Water enters the fish's mouth, is forced over the gills, and then exits the fish's body. As the water passes over the gills, oxygen moves from the water into the fish's blood stream.

If a fish swims with its mouth open, there is a constant flow of water through the gills. If a fish is not swimming, it can force the water through the gills by contracting its throat. Some sharks cannot do this so they must continually stay in motion.

Fish School:

A group of fish is called a school just like a group of birds is called a flock. Have your child create an underwater picture. Try to include seaweed, sand, rocks or other sea life. Then glue a group of goldfish snack crackers on the picture to show a school of fish. Take a picture of your creation to include in your animal notebook. It can also be fun to eat some of the fish.

Name a Group Game:

Have one child name an animal, then have another child name what a group of that type of animal would be called. Here are a few to get you started:

A school of fish, a flock of birds, a gaggle of geese, a herd of elephants, a pride of lions, a pack of wild dogs, a flock of sheep, a brood for vipers, a swarm of flies

What did we learn?

What makes fish different from other animals? (They live in the water, are cold-blooded, have gills and scales.)

How do fish breathe? (They get oxygen from the water using gills.)

Why do some sharks have to stay in motion? (They must have a constant flow of water over their gills to breathe and the only way they can do that is to swim with their mouths open.)

Taking it further

Other than how they breathe, how are dolphins different from fish? (Dolphins are warm-blooded, give live birth, nurse their young, have hair, and do not have scales.)

How are dolphins like fish? (They live in the water, swim, have fins and a tail, and eat fish.)

Fins and Other Fish Anatomy

Designed for efficiency

Supply list:

- 1 copy of "Fish Fins" worksheet for each child (pg. 63)

- Construction paper

- Scissors

- Glue

Most fish have a similar body shape. God designed a fish's body to be very efficient in the water. First, gills are designed to remove up to 80% of the oxygen from the water. By comparison, lungs usually remove only about 25% of the oxygen from the air.

To be efficient swimmers, God designed fish with a long narrow body that glides easily through the water. The fish's body produces slimy mucus that coats the body and helps it swim more easily. To stay afloat, most fish have a swim bladder. This is a balloon-like sac that can be inflated with air to help the fish rise in the water, or it can be deflated to help the fish sink or go deeper in the water.

Finally, fish were designed with several different fins to help them be great swimmers. Fish have two pairs of fins toward the front of their bodies. Pectoral fins are located on the sides behind the mouth. The pelvic fins are lower down on the body. Pectoral and pelvic fins allow the fish to angle up or down when swimming. They can act as brakes to slow the fish down. And they can even be used to slowly move the fish backward. These fins also help the fish swim in a straight line.

Fish usually have one or two fins that stick up on their backs. These are called dorsal fins. They also have a fin pointing down from the bottom called an anal fin. Dorsal and anal fins keep the fish from tipping sideways and give it stability.

The final fin is the fish's tail called the caudal fin. The caudal fin is the fish's rudder, which is used for steering.

All of these designs, along with strong muscles, make the fish one of the most efficient swimmers in creation. God's great design is very evident in the fish's body.

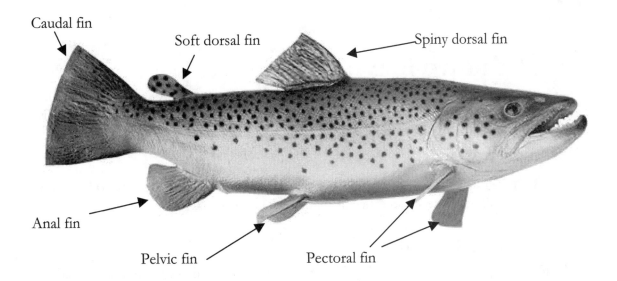

Caudal fin

Soft dorsal fin

Spiny dorsal fin

Anal fin

Pelvic fin

Pectoral fin

One of the few exceptions to the fast swimming fish is the seahorse. This tiny creature may not even seem like a fish. But it is cold-blooded, has fins, scales, and gills, and lays eggs like other fish. This tiny fish has a dorsal fin that moves it forward in an upright position at a rather slow speed. However, just because it is slow does not mean it was not well designed. God designed the seahorse to fit into a different food chain than most other fish and equipped it to survive in its environment.

Fins Worksheet:

Have your child cut fins from construction paper and glue them in the correct places on the "Fish Fins" worksheet. Have your child label the fins and color the fish. This page can be included in the animal kingdom notebook.

What did we learn?

What is the purpose of a swim bladder? (It gives the fish buoyancy. When it fills with air it makes the fish lighter than the water, allowing it to rise. When the air is released the fish becomes heavier than the water and it sinks. This buoyancy keeps the fish floating without having to keep moving its fins.)

How did God design the fish to be such a good swimmer? (The shape of its body, its fins, and the mucus all help it to be an efficient swimmer.)

Taking it further

How does mucus make a fish a more efficient swimmer? (Since mucus is slippery, it reduces friction so the fish does not have to work as hard to move through the water. To see how this works, put your hand under some running water and watch how the water flows. Then, rub a little cooking oil on your hand and repeat the test. The water flows more quickly over your oily hand because there is less friction.)

How has man used the idea of a swim bladder in his inventions? (Submarines use air to help keep them afloat at the desired depth. Also, life rafts fill with air to help them float to the top of the water.)

What other function can fins have besides helping with swimming? (Fins can provide protection from predators. Fins can make it difficult for a predator to swallow a fish. In addition, some fins are shaped and colored to help provide camouflage.)

What similar design did God give to both fish and birds to help them get where they are going? (They both have rudder-like tails that help them steer, and bodies specially shaped for moving through their environments.)

Fish Fins

The fish below is missing its fins. Add fins to make the fish a good swimmer. The list below should help you know what fins to add.

Pectoral fins
Pelvic fins
Dorsal fin
Anal fin
Caudal fin (already drawn)

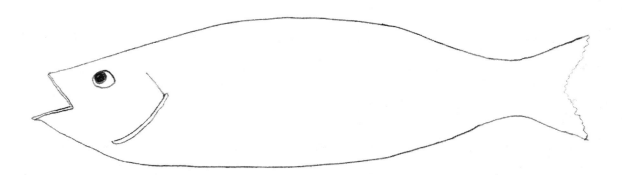

Cartilaginous Fish

No bones about it!

Supply list:

- Modeling clay

About 5% of all fish do not have bony skeletons. These fish have cartilage structures instead. Cartilage is a tough yet flexible material that provides shape and structure without being stiff. Cartilage is what gives your nose and ears their shape. The most well-known cartilaginous fish is the shark. These fearsome fish can be as small as 6 inches or as big as 60 feet long.

Sharks are meat-eating creatures that can be found in most parts of the ocean. Most sharks have several rows of razor sharp teeth. When one row of teeth wears out, the next row moves up to take its place.

Most sharks have the same general shape as the bony fish. However, their internal structure and function are quite different. First, sharks do not have swim bladders. If sharks quit swimming they sink to the ocean floor. Also, many sharks cannot force water out of their throats, so they must remain in constant motion to keep water flowing over their gills. Sharks do not have covers over their gills. Instead, they have gill slits that are easily visible on the sides of their bodies. Finally, many sharks give live birth instead of laying eggs.

Shark babies are born ready to care for themselves and leave their mothers right away. Sharks that lay eggs usually place their eggs in a thick egg sac and then never return. The young are on their own when they hatch.

Another type of cartilaginous fish is the ray. Rays do not have slim tapered bodies; instead they are wide and flat. They glide through the water using their dorsal fins like wings. Rays are generally harmless to man -- except for the stingray. The stingray has a whip-like tail that can inject venom into its enemies or cause a painful sting in humans.

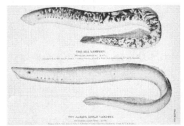

Other cartilaginous fish that are not shaped like most fish are the lampreys and hagfish. These fish resemble snakes with their long round bodies. They do not have jaws like other fish. Instead, they have a sucking type of mouth. Lampreys are parasites, animals that take nutrients from a living host, often harming the host, and attach themselves to other fish and suck nutrients from their bodies. Lampreys are similar to eels. However, eels have bony skeletons and jaws, while lampreys have cartilaginous skeletons and no jaws. Hagfish are generally scavengers, animals that eats dead plants or animals.

Clay models:

Have your child use modeling clay to make models of cartilaginous fish such as sharks and rays. When he is done, take a picture to put in his animal kingdom notebook.

What did we learn?

How do cartilaginous fish differ from bony fish? (Their skeletons are made from cartilage instead of bone. Also, many of these fish do not have the typical torpedo shaped body.)

Why is a lamprey called a parasite? (It does not eat prey. Instead, it attaches its mouth to a living animal, usually a fish, and sucks its blood for nutrients.)

Why can sharks and stingrays be dangerous to humans? (Sharks can attack with their sharp teeth and stingrays can sting with their tails.)

Taking it further

Why are shark babies born independent? (Like many other animals, sharks do not care for their young, so the babies must be able to care for themselves at birth. Many adult sharks will eat young sharks so babies must avoid adults until they are large enough to defend themselves.)

What do you think is the shark's biggest natural enemy? (Other sharks)

Vertebrates Quiz

Lessons 2-19

What defines an animal as a vertebrate? _____

_____

Place the letters of the characteristics that apply next to each animal group.

Mammals _____   A. Warm-blooded

                                 B. Cold-blooded

Birds _____   C. Has hair or fur

                                 D. Has scales

Fish _____   E. Has feathers

                                 F. Has no hair, feathers, or scales on its skin

Reptiles _____   G. Gives live birth

                                 H. Lays eggs

Amphibians _____   I. Experiences metamorphosis

                                 J. Has lungs

                                 K. Has gills

                                 L. Nurses its young

                                 M. Has wings

                                 N. Has fins

(Note: answers to all quizzes and tests are in Appendix A)

Invertebrates

Creatures without a backbone

Supply list:

- A good imagination

- Optional: white board with markers

Although most familiar animals are vertebrates, the vast majority (nearly 97%) of all animals are invertebrates: animals without a backbone. Invertebrates do not have internal skeletons; therefore, most of them are small creatures. The squid and the octopus are the only large invertebrates.

The huge variety among invertebrates makes it difficult to group them together. However, scientists have grouped most invertebrates into six groups: arthropods, mollusks, coelenterates, echinoderms, sponges, and worms. In these groups are familiar creatures such as jellyfish, corals, starfish, crabs, shrimp, spiders, ladybugs, and butterflies.

Invertebrates can be found in every part of the world. Some may be found in your home, such as spiders and insects. Perhaps you had invertebrates for dinner. You did if you ate shrimp or clams. Earthworms live in your garden. Many more invertebrates live in the waters of rivers, lakes, and oceans. Many invertebrates are microscopic and can even be found inside of you, helping to digest your food.

Invertebrates are an indispensable part of the food chain. They feed the largest creatures of the sea. Whales eat tons of small invertebrates each day. Also, invertebrates help to break down dead plant and animal tissue to be recycled. We will find God's marvelous creativity all around us as we explore the world of invertebrates.

Invertebrate Repeat Game:

This is a simple memory game. Have the first person name an invertebrate. The second person repeats that animal and names a new and different invertebrate. The third person says the first 2 animals and adds a new one. The game continues until someone breaks the chain or cannot name a new invertebrate.

To help your children recognize invertebrates, you can review the vertebrates: mammals, birds, amphibians, reptiles, and fish. Make sure they understand that they are not to name any of these animals in the game. Then help them think of a few invertebrates to get started.

This can be a lot of fun, and can help you and your children realize that invertebrates are really more common than they thought.

Optional: Invertebrate Pictionary:

Have one child draw a picture of an invertebrate on a white board. While he is drawing have the other children guess what the animal is. The first person to guess the correct animal gets to be the artist.

What did we learn?

What are some differences between vertebrates and invertebrates? (The main difference is that vertebrates have backbones that protect their spinal cords. Invertebrates do not have backbones or spinal cords. Also, vertebrates have internal skeletons and invertebrates don't.)

What are the six categories of invertebrates? (Arthropods, mollusks, coelenterates, echinoderms, sponges, and worms)

Taking it further

Why might we think that there are more vertebrates than invertebrates in the world? (We don't notice invertebrates as much as we do vertebrates. Invertebrates are usually small, a great many are microscopic, and so we just don't see them as often. Also, many invertebrates live in the water, so, again, we don't see them very often.)

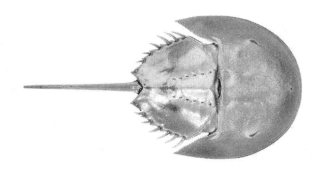

Arthropods

Invertebrates with jointed feet

Supply list:

- 1 copy of the "Arthropod Pie Chart" for each child (pg. 71)

The largest group of invertebrates is the arthropods. Over 75% of all animal species are arthropods. Arthropod means "jointed foot" so obviously all of the creatures in this group have jointed feet or jointed legs. In addition, all arthropods have segmented bodies, meaning they have two or more distinct body regions. Arthropods also have exoskeletons. Instead of internal bones or cartilage like vertebrates, arthropods get their protection and structure from an external covering or exoskeleton. This outside skeleton is both strong and flexible. It is made of chitin, a starchy substance.

Of the five groups of arthropods, the largest group is insects (90%). The other groups are arachnids (6%), crustaceans (3%), millipedes (0.8%), and centipedes (0.2%). There are approximately 1 million different insects that have been identified, and scientists believe there may be as many as 1 million more species that have not yet been classified. Some of the more common insects include flies,  beetles, mosquitoes, butterflies, and ants. Arachnids include spiders, ticks, mites, and scorpions. Nearly all crustaceans live in the sea, and include shrimp, crabs, and lobsters. Centipedes and millipedes are similar creatures, both having multiple body segments and many legs.

As you study this large diverse group of invertebrates, look for evidence of God's design.

Arthropod Pie Chart:

Use the following information to label the "Arthropod Pie Chart."

Insects:	1,000,000 species
Arachnids:	70,000 species
Crustaceans:	30,000 species
Millipedes:	10,000 species
Centipedes:	2,800 species

Include this pie chart in the animal kingdom notebook.

What did we learn?

What do all arthropods have in common? (They are invertebrates (no backbones) with jointed feet, segmented bodies, and exoskeletons.)

What is the largest group of arthropods? (Insects, with over 1 million species)

Taking it further

How are endoskeletons (internal) and exoskeletons (external) similar? (They both provide support and protection for the body. They help give the animal its form and shape.)

How are endoskeletons and exoskeletons different? (Endoskeletons are on the inside of the body and are usually made of bone or cartilage. Also, endoskeletons grow as the body grows. Exoskeletons are on the outside of the body and are made from chitin, a substance similar to starch. Exoskeletons do not grow with the animal and must be shed periodically as the body grows.)

Why should you be cautious when hunting for arthropods? (Many arthropods are poisonous, including some spiders, scorpions, and centipedes.)

Arthropod Pie Chart

Use the words from the box to label the chart.

| Insects | Crustaceans | Arachnids | Centipedes | Millipedes |

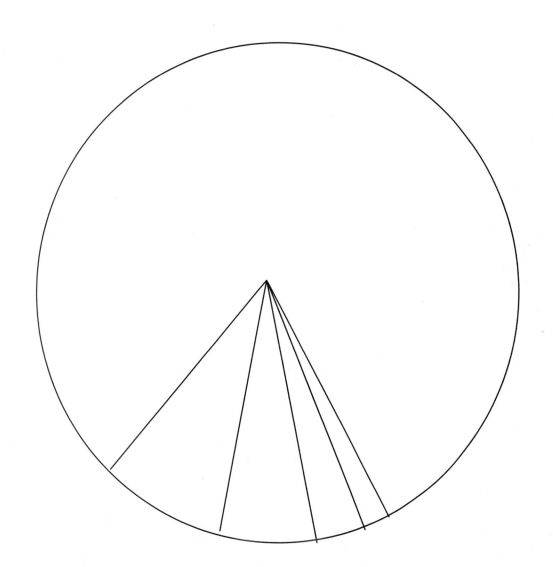

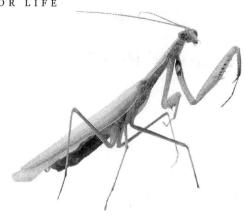

Insects

Don't let them bug you.

Supply list:

- 3 styrofoam balls per child
- 4 pipe cleaners per child
- 2 toothpicks per child
- Paper
- Scissors
- Tape
- 1 copy of "Water Skipper Pattern" (pg. 74)

- Paint
- 1 index card per child
- Bowl of water

There are over 1 million kinds of insects. No wonder you have trouble keeping them out of your house. Insects are the largest group of arthropods. In addition to jointed legs and exoskeletons, insects have 3 distinct body parts: the head, thorax, and abdomen. They also have a pair of antennae on their heads, as well as simple and compound eyes. Insects have 6 legs attached to their thoraxes and most have 1 or 2 pairs of wings.

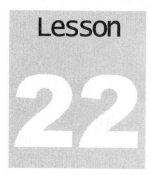

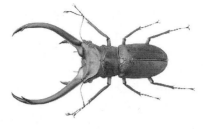

Because there are so many different types of insects, scientists have grouped them into categories by similar characteristics. One group has straight wings. This includes grasshoppers and crickets. Half-wings are the true bugs such as the stinkbug. Butterflies and moths are in their own group. Flies and mosquitoes are in another. Beetles include the stag beetle, weevil, and June bug. There are many other groups of insects as well.

Many insects are pests. They can destroy crops and spread diseases. Insects can cause painful bites, and just get annoying. However, insects play a very important role in the ecosystem. Birds, reptiles, amphibians, and many other animals depend on insects for food. Insects are also very important for pollinating flowers. And some insects, such as butterflies, are very pleasant to have around. As much as we might like to get rid of them, insects are vitally needed.

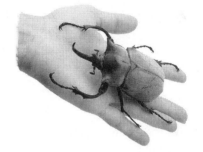

Insect Models:

To make an insect model:

Connect 3 styrofoam balls together with toothpicks to form the body of the insect. If desired, paint the balls the color of the insect; for example, paint it yellow and black if it is a bee or red or black if it is an ant. While the paint is drying, cut pipe cleaners into eight equal length pieces. Use six of these pieces as legs. Insert the legs into the center ball (the thorax), three on each side, and bend them to look like legs. Insert the other two pieces of pipe cleaner into the head for antennae. Cut two or four wings from paper. Use brightly colored paper if you are making a butterfly. Many other insects have translucent wings so you could use white paper or make a frame from pipe cleaners and cover the frame with plastic wrap. If using paper, tape the wing pieces to additional pieces of pipe cleaner and insert them into the center ball.

Discuss each part of the insect as you put it together. Take pictures of the models and include them in the animal notebook.

To make a water skipper model:

Fold an index card in half with long sides together. Trace the pattern below onto the card and cut out the bug shape. Fold the feet out. Place the card in a bowl of water very carefully. The card should float on the surface of the water. Discuss how water molecules are attracted to each other so things that are light enough can stay on the surface. Water skippers and other light insects can walk across the surface of the water without sinking because they are light enough not to break the surface tension.

Water Skipper Pattern

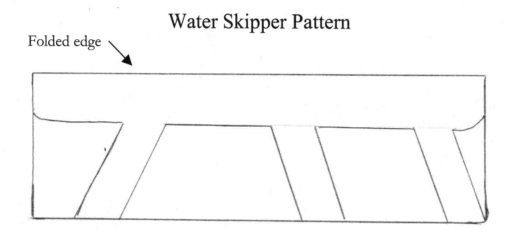

Folded edge

Optional Activity:

Playing the game "Cootie" from Milton Bradley is a fun way to review the parts of an insect if you have it available.

What did we learn?

What characteristics classify an animal as an insect? (Insects are invertebrates with jointed feet, three body parts (head, thorax, and abdomen), 6 legs, antennae, and usually have wings.)

How can insects be harmful to humans? (They can destroy crops and spread disease.)

How can insects be helpful? (Some insects eat other insects. For example dragonflies eat mosquitoes and ladybugs eat aphids. Many insects pollinate flowers. Other insects provide food for many other animals.)

Taking it further

How might insects make noise? (Some insects make noise by flapping their wings. Others, like crickets, can rub their legs together to make noises.)

Insect Metamorphosis

Making a change

Supply list:

- 1 copy of "Stages of Metamorphosis" worksheet for each child (pg. 78)

- Sleeping bag

- 2 scarves or other pieces of cloth

- Optional: Butterfly larvae and habitat

All insects reproduce by laying eggs. However, what hatches out of the egg may or may not look anything like the parents. Most insects go through a metamorphosis or change between birth and adulthood.

Some insects, such as grasshoppers, go through incomplete metamorphosis. This means the young resemble their parents when they hatch and gradually change into an adult. There are three stages to an insect's life if it experiences incomplete metamorphosis: egg, nymph, and adult. The nymph hatches from the egg, then as it grows it molts or sheds its exoskeleton several times. As it gets bigger the nymph begins growing wing pads. After its final molt, the insect has complete wings and is considered an adult. Dragonflies, crickets, termites, and grasshoppers are some of the insects that experience incomplete metamorphosis.

Egg   Two stages of nymph   Adult

Most insects experience a 4-stage lifecycle called complete metamorphosis. An insect that goes through complete metamorphosis starts out as an egg. When it hatches it is called a larva. The larva of the butterfly is a caterpillar. Other insect larvae also resemble caterpillars. The larva does not look much like the adult that it will become.

As the larva grows, it spends most of its time eating. Its exoskeleton cannot grow with it, so it sheds the exoskeleton several times as it grows. After a few days or weeks the larva enters the third stage of its life, called the pupa or chrysalis stage. During this stage the larva's body undergoes a tremendous change. This stage can last from a few days to a few months, depending on the type of insect.

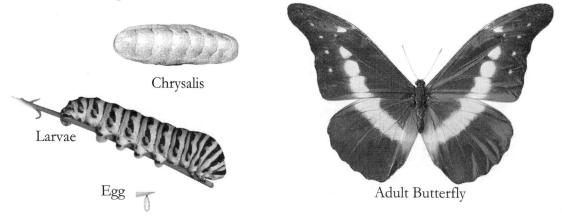

Chrysalis

Larvae

Egg

Adult Butterfly

When the change is complete, the insect breaks out of its chrysalis and is an adult. The most dramatic change to witness is the changing of a caterpillar into a butterfly, but other insects experience drastic changes as well. Witnessing this change is an amazing experience.

Metamorphosis Worksheet:

Have your child draw and color the different stages of metamorphosis on the "Stages of Metamorphosis" worksheet. Include this worksheet in the animal notebook.

Butterfly Breakout:

Place two scarves inside a sleeping bag. Have your child pretend to be a caterpillar and crawl into the sleeping bag. He should remain very still while in the sleeping bag, just as the pupa does not move inside the chrysalis. Then, have him unzip the sleeping bag and emerge holding the scarves and using them like wings. He has become a butterfly.

Optional – Observe Metamorphosis:

The very best way to appreciate the changes that occur in complete metamorphosis is to actually observe them. Many science supply catalogs sell live caterpillars that can be kept and observed while they turn into butterflies. (See Appendix B for possible suppliers.) The butterflies can then be released if the weather is favorable.

What did we learn?

What are the three stages of incomplete metamorphosis? (Egg, nymph, adult)

What are the four stages of complete metamorphosis? (Egg, larva, pupa (or chrysalis), adult)

Taking it further

What must an adult insect look for when trying to find a place to lay her eggs? (The eggs must be laid on a plant that the larva can eat. A larva spends most of its time eating and cannot search for food, so food must be readily available.)

FUN FACT

Monarch butterflies are one of the few insects that migrate with the changing seasons. Huge groups of monarchs can sometimes pass through an area for several days or even weeks as they migrate.

Stages of Incomplete Metamorphosis

Egg   Nymph   Adult

Stages of Complete Metamorphosis

Egg   Larva

Pupa   Adult

Lesson

24

Arachnids

Spiders and such

Supply list:

- Several large and small marshmallows for each child

- Flexible wire, about 4 inches for each child.

- 4-6 pipe cleaners per child

- 1 toothpick per child

Optional:

- 2 round crackers per child

- Peanut butter

- 2 raisins per child

- 8 pretzel sticks per child

Many people wonder if spiders are insects. Despite some similarities in their looks, there are several differences between insects and spiders. Insects have three body parts, whereas spiders have only two body parts (called the cephalothorax and the abdomen). Also, insects have six legs and spiders have eight. These differences place spiders in the class of arachnids. In addition to two body parts and eight legs, arachnids also lack the wings and antennae commonly found in insects.

Spiders are the most common arachnids. They can be found throughout the world. Most spiders spin webs and kill their prey with venom. But only a few spiders are poisonous to humans. Spiders do not have mouth parts for biting or chewing. They can only suck liquids from their prey. Spiders have special organs at the back of their abdomen called spinnerets that produce the silk thread used in weaving webs. If you closely examine a spider's web, you will find that some of the strands are smooth and some are sticky. The spider can move easily around its web by walking on the smooth strands. Also, spiders secrete an oily substance on their feet that keeps them from sticking to their own webs.

Spiders are not the only creatures in the arachnid family. Mites, ticks, and scorpions are also arachnids. Mites and ticks resemble small spiders. However, they do not spin webs or catch insects. Instead, they attach themselves to other animals and suck their blood. For many creatures mites and ticks are a nuisance. But for some, including humans, mites and ticks can carry and spread serious diseases. That is why it is always a good idea to check for ticks after hiking in the woods or other areas where ticks live.

At first glance, scorpions may not seem to fit in with spiders, ticks, and mites. But a close examination shows that scorpions have 8 legs attached to a cephalothorax as well as an abdomen. The scorpion's abdomen is jointed and curls behind it, ending in a stinger. Scorpions can inflict a painful sting and should be avoided.

Spider and Scorpion Models:

To make a spider model:

Use two large marshmallows as the two body parts. Connect them with a toothpick. Cut two pipe cleaners into four pieces each. Insert them into the front marshmallow (four on each side). Bend the pipe cleaners to resemble legs. You can draw eyes on the front of the spider with marker if you want to. Discuss the different parts with your child as he assembles the model.

To make a scorpion model:

Use one large marshmallow for the cephalothorax (front) and several small marshmallows for the abdomen (tail). Using a piece of flexible wire, string several small marshmallows together and attach them to the end of the large marshmallow. Bend the wire up so the small marshmallows resemble a tail with a stinger. Use pipe cleaners to make eight legs and stick them into the large marshmallow. Twist a small piece of pipe cleaner onto the end of each of the front legs to make claws. Remind your child that scorpions are poisonous.

Take pictures of the models to add to the animal kingdom notebook.

Optional – Spider Snacks:

For a fun snack you can make edible spider snacks. Use a round cracker for the cephalothorax. Spread peanut butter on the cracker. Add raisins for eyes. Add 8 pretzels sticks to the cephalothorax for legs. Add another round cracker to one edge for the abdomen. Spread peanut butter on this cracker as well. Yum!

Optional – Looking at a Web:

If you have the opportunity, closely observe a real spider's web. Use a magnifying glass to look at the individual strands. Sprinkle a light powder, such as powdered sugar, on the web. It will stick to the sticky strands but not to the non-sticky ones.

What did we learn?

How do arachnids differ from insects? (They have only two body parts (cephalothorax and abdomen), eight legs, no wings or antennae, and many spin webs.)

Why are ticks and mites called parasites? (They feed off of living hosts.)

Why don't spiders get caught in their own webs? (Only some of the web strands are sticky. The spider walks on the ones that are not sticky. Also, spiders secrete an oily substance that coats their feet and keeps them from sticking to their webs.)

Taking it further

At first glance, scorpions and crayfish look a lot alike. How does a scorpion differ from a crayfish? (A scorpion lives on land and has eight legs, a stinger, and no antennae. Crayfish live in the water and have ten legs, antennae, and no stingers.)

FUN FACT

Daddy longlegs are not actually spiders. Although they look very similar to spiders, they do not have spinnerets. Daddy longlegs have only one pair of eyes, whereas spiders have eight eyes. Also, spiders suck liquids from animals, but daddy longlegs suck liquids from plants as well as from animals.

Crustaceans

Are they crusty?

Supply list:

- Modeling clay

- Magnifying glass for optional activity

H ave you ever chased a crawdad in a stream or eaten shrimp or lobster? Have you ever watched a crab burrow in the sand or dug up a roly-poly? If so, then you are familiar with crustaceans. Most crustaceans live in the water with the notable exception of the pill bug (often called a roly-poly).

All crustaceans are arthropods with jointed legs and exoskeletons. In addition, crustaceans have 2 distinct body parts or regions: the cephalothorax or head region and the abdomen. They also have two pairs of antennae, two or more pairs of legs, and gills for breathing.

The crawdad or crayfish is a very familiar crustacean found in many fresh water streams. Crayfish have two pairs of antennae on their heads that help them sense food or enemies. They also have 5 pairs of legs attached to their cephalothoraxes. The front legs end in pincers or claws, which they use for catching prey or defending themselves. Because a crayfish eats food from the bottom of the riverbed, God put its mouth on the underside of its body. Also, a crayfish can evade an enemy by darting backward very quickly.

You may be most familiar with the larger crustaceans such as crabs, shrimp, lobsters, and crayfish. But the majority of crustaceans are very small – mostly microscopic. Brine shrimp, water fleas, and other tiny crustaceans are very important links in the aquatic food chains. Many sea creatures depend on these tiny creatures for food. Many whales have special "teeth" called baleen, which they use to strain these tiny crustaceans from the water.

Clay Models:

Have your child make clay models of several different crustaceans. Point out the similarities and differences between the animals. Take pictures of the models to add to the animal kingdom notebook.

Optional Activity - Roly-poly Exploration:

If weather permits, search for a roly-poly. Because they have gills, they must stay in moist areas so they can usually be found under rocks, pieces of wood or in other protected areas. If you find one, examine it carefully with a magnifying glass. Pay close attention to its jointed legs, antennae, exoskeleton, and segmented body.

What did we learn?

What do all crustaceans have in common? (They have jointed legs, exoskeletons, two body sections, two pairs of antennae, two or more pairs of legs, and gills.)

How did God specially design the crayfish for its environment? (It was designed with claws for defense and for eating. Its mouth is on the underside of its body, making it easier to eat food from the bottom of the river.)

Taking it further

Why might darting backward be a good defense for the crayfish? (It is unexpected and can confuse an enemy.)

How can something as large as a blue whale survive by eating only tiny crustaceans? (It eat lots and lots of them – up to 8000 pounds per day!)

If you want to observe crustaceans, what equipment might you need? (Jar, net, microscope, trap)

Myriapods

How many shoes would a centipede have to buy?

Supply list:

- A good memory and a sense of adventure

- (A baseball cap might be fun too.)

Centipedes and millipedes are called myriapods (meaning "many feet"). Centipede means 100 feet and millipede means 1000 feet but do centipedes and millipedes really have that many feet? No, but they do have a lot of feet. Let's see just how many feet they have.

Centipedes have between 15 and 25 body segments. Each segment has 1 pair of legs, so a centipede has between 30 and 50 legs. Its first pair of legs has poisonous claws that are used to kill its prey. Centipedes also have long antennae and a flattened body that is usually a few inches long.

Millipedes differ from centipedes in several ways. First, they have rounded bodies, not flat bodies. Also, they are not poisonous. But the most distinctive difference is that they have  two pairs of legs per body segment instead of only one pair. Millipedes can have between 44 and 400 feet but not 1000. Millipedes tend to be bigger and slower than centipedes and have shorter antennae. Both centipedes and millipedes live in dark moist places.

Centipedes and millipedes are sometimes confused with caterpillars. However, caterpillars do not have legs on every body segment. Also, caterpillars experience metamorphosis and change into butterflies or moths. Centipedes and millipedes do not change into another form. Caterpillars are often fuzzy and live in the open on different plants. Myriapods are usually smooth and live in dark places such as under rocks or under ground.

Arthropod Baseball:

Set up a "baseball diamond" by assigning places such as chairs or pieces of paper on the floor to be home, 1st, 2nd, and 3rd base. Set a chair in the middle as the pitcher's mound. Mom gets to be the pitcher. Each child gets to be a batter. The batter selects the difficulty of the pitch: single, double, triple, or home run. The pitcher selects an appropriate question about arthropods. For example, "Name the 5 groups of arthropods." The batter must answer the question. If the answer is correct, the batter advances to the appropriate base, (i.e. 1 base for a single, 2 bases for a double. The harder the question, the more bases the question is worth.) Then the next batter gets to answer a question.

If the batter cannot correctly answer a question, it is considered an out. After three outs, Mom gets to be the batter and the kids get to ask the questions. See who can score the most runs.

What did we learn?

How can you tell a centipede from a millipede? (Centipedes are usually smaller, flatter, and have longer antennae. Also, centipedes have 1 pair of legs per body segment and millipedes have 2 pairs per segment.)

What are the 5 groups of arthropods? (Insects, arachnids, crustaceans, centipedes, and millipedes)

What do all arthropods have in common? (They have jointed legs, exoskeletons, and 2 or more body regions.)

Taking it further

What are some common places you might find arthropods? (Nearly everywhere!)

Arthropods are supposed to live outside, but sometimes they get into our homes. What arthropods have you seen in your home? (Probable answers include: ants, flies, mosquitoes, and spiders.)

Arthropods Quiz

Lessons 21-26

Write Yes if the creature below is an arthropod, write No if it is not.

1. _____ ant   2. _____ tick   3. _____ trout   4. _____ spider

5. _____ scorpion   6. _____ crab   7. _____ cricket   8. _____ butterfly

9. _____ clam   10. _____ mouse   11. _____ centipede   12. _____ snail

13. _____ roly-poly   14. _____ crawdad   15. _____ starfish   16. _____ lizard

17. What are the four stages of complete metamorphosis for an insect?

_____ _____ _____ _____

Fill in the blanks with the appropriate numbers:

18. An insect has _____ body parts, _____legs, _____ antennae, and _____ wings.

19. A spider has _____ body parts, _____ legs, _____antennae, and _____ wings.

20. A centipede has _____pair(s) of legs per body segment and a millipede has _____ pair(s) of legs per body segment.

(Note: answers to all quizzes and tests are in Appendix A)

Mollusks

Creatures with shells

Supply list:

- Several sea shells

- Shell identification guide

As you walk along the beach you are likely to find a variety of sea shells. Most of these shells are the remains of mollusks. Mollusks are soft-bodied invertebrates. They have non-segmented bodies with no bones. Mollusks have one muscular foot for moving about, a hump containing the internal organs, and a mantle, which is an organ that secretes a substance that hardens into a shell in most species. Most, but not all, mollusks live in the water.

Although all mollusks have these characteristics, there is great variety among them. Mollusks that have two-part shells are called bivalves. Oysters, clams, scallops, and mussels are all bivalves. The shells of these mollusks are connected by a hinge at the back and are opened and closed by strong muscles. Many bivalves produce a pearly substance that protects the internal organs from irritants that get inside their shells. Oysters produce a shiny substance that, after a period of years, turns the irritants into pearls that are prized by people.

Many mollusks have only one-piece shells. These are called gastropods. Gastropods include snails, conchs, abalones, and slugs. With the exception of slugs, gastropods produce beautifully spiraled shells (like the one at the top of this page). Each species produces a unique style of shell, so shells can be used to identify the animal.

The third group of mollusks is the cephalopods. This group includes squids, octopuses, and nautiluses. At first glance, this group may not seem to fit the characteristics of mollusks. However, a closer examination reveals a foot merged with the head, a hump containing the internal organs, and a mantle. In many cephalopods the mantle produces an outer body wall and not a rigid shell.

Squids and octopuses can move quickly through the water by jet propulsion. Also, both creatures can spray out an inky substance when they feel threatened to confuse their enemies and get away. Both squids and octopuses have complex eyes that are similar in designed to the eyes of vertebrates. Evolutionists have a very hard time explaining how creatures with such supposedly different evolutionary chains as octopuses and humans, ended up with such similar and complex eyes. The giant squid that lives in the Pacific and North Atlantic Oceans can grow to be 60 feet long, weigh up to 1 ton, and is the largest invertebrate. Octopuses have a complex brain and are considered to be the most intelligent invertebrates.

Shell Identification:

Collect as many different shells as you can. Each mollusk generates a unique shape of shell; therefore, you can tell what animal used to live in it if you have a guide to help you. Use a shell guidebook to help you identify what creature used to live in each of your shells. Which shells were bivalves? (Two parts)   Which shells were gastropods? (One part – spiraled) Take pictures of your shells and include the pictures and identifications in your notebook.

What did we learn?

What are 3 groups of mollusks? (Bivalves, gastropods, and cephalopods)

What body structures do all mollusks have? (They all have soft bodies, a muscular foot, a hump for internal organs, and a mantle that forms a shell in most species.)

How can you use a shell to help identify an animal? (The size, shape, and coloring of each shell are unique to its species. Some shells spiral counter-clockwise and others spiral clockwise. Some are two pieces and some are only one piece.)

Taking it further

How are pearls formed? (Any irritant that gets inside an oyster's shell is coated with a pearly substance over and over again. After a period of several years, it is large enough to be of value to people. To speed up this process, many oyster farmers now "seed" oysters by placing hard round objects that are nearly the size of a pearl inside oyster shells. After only a few months these artificial pearls are ready for harvesting.)

Coelenterates

Jellyfish, coral, and sea anemones

Supply list:

- Paper

- Modeling clay

- Marshmallows

- Pretzel sticks

- Whatever other craft supplies you have on hand

- Pictures of a coral reef

What do jellyfish, coral, and sea anemones have in common? They all have hollow bodies and stinging tentacles, so they are grouped together as coelenterates. Coelenterate (see-LEN-ter-ate) means "hollow intestine." Coelenterates spend at least part of their lives in the form of a polyp. A polyp has a cylinder-shaped body with tentacles and a mouth on top.

This description of a polyp may not make you think of a jellyfish, but the jellyfish goes through many changes in its lifecycle. It begins life as an egg. The egg hatches into a planula, which looks like a worm. The planula settles on the sea bottom and grows into a polyp. It may remain in the polyp stage for up to a year. Eventually disks grow at the top of the polyp. When these disks break off they become medusas. When the medusas are mature they have the jelly-filled bodies we consider to be adult jellyfish. Jellyfish inhabit nearly all parts of the ocean. Jellyfish have stinging tentacles, so most creatures stay away from them. A few animals however, have special protection from the stings and can live closely with jellyfish.

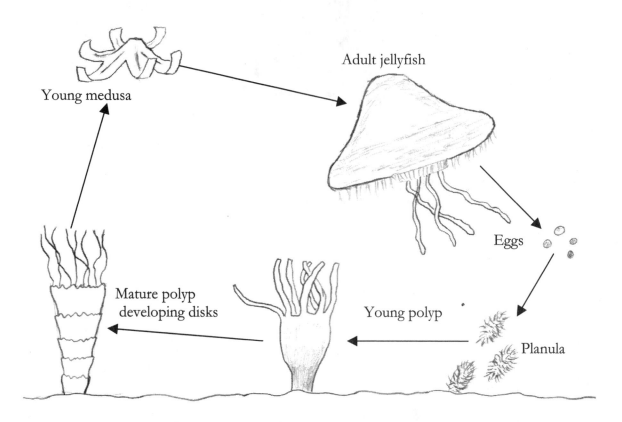

Young medusa

Adult jellyfish

Eggs

Mature polyp
developing disks

Young polyp

Planula

Coral may not appear to be related to jellyfish but their design is very similar to the polyp stage of the jellyfish. A coral also begins life as an egg that hatches into a larva. The larva settles on the sea bottom and begins secreting a calcium-based substance. This substance hardens into a case around the larva. The larva then develops into a polyp. Hundreds of thousands of polyps live together to form a coral colony. Millions of coral colonies create a coral reef. Coral polyps stay inside their hard cases when threatened, but when hunting, the polyps come out and shoot poison arrows at their prey. Then the polyps' tentacles pull the stunned food into their hollow bodies.

Sea anemones have a similar polyp body structure to jellyfish and coral. They are often brightly colored and catch their prey with their long stinging tentacles.

Several coelenterates have symbiotic or mutually beneficial relationships with other creatures. For example, some corals have an algae living inside them. This algae produces food for the coral, while the coral provides protection for the algae. Also, some fish that are not harmed by jellyfish live near them. When a larger fish attacks the smaller fish, the jellyfish stings the attacker, then both the jellyfish and the smaller fish share the meal.

Coral Reef:

Make a model of a coral reef. Look at pictures of different kinds of coral. Then use clay, marshmallows, pretzel sticks, or any other craft supplies you have available. Glue these items together on a cardboard base to make a model of a coral reef. Take a picture of your reef and include it in your animal kingdom notebook.

What did we learn?

What characteristics do all coelenterates share? (They have hollow bodies with stinging tentacles.)

What are the three most common coelenterates? (Jellyfish, coral, and sea anemones.)

Taking it further

How do you think some creatures are able to live closely with jellyfish? (Some animals have a tough skin or exoskeleton that protects them from jellyfish stings. Others have a special coating on their skins that protects them.)

Why do you think an adult jellyfish is called a medusa? (The Medusa was a mythological creature with snakes for hair. A jellyfish, with all of its tentacles, resembles this creature.)

Jellyfish and coral sometimes have symbiotic relationships with other creatures. What other symbiotic relationships can you name? (Some birds eat insects off of cattle. This feeds the birds and helps the cattle stay healthy. Also, lichen, that green and yellow scaly-looking substance on rocks, is actually fungus and algae living in a symbiotic relationship. The algae have chlorophyll and produce the food, while the fungus provides water, nutrients, and protection. It is a beneficial relationship for both organisms.)

Lesson

29

Echinoderms

Spiny skinned creatures

Supply list:

- Salt dough (1 cup salt, 1 cup flour, water to make a stiff dough)

- Tag board or cardboard

- Mini-chocolate chips

- If possible: a real (dead) starfish or sand dollar (sometimes available at craft stores)

Echinoderms (ee-KINE-o-derms) are spiny-skinned animals. These creatures have hard spikes made from calcium carbonate. Echinoderms also have a system of water-filled tubes that help them move. The most familiar echinoderms are starfish, sea urchins, and sand dollars. Most echinoderms have a central disk with 5 rays going out from the disk. This design is easily seen in starfish but can be observed in sand dollars as well.

Starfish are the most well known echinoderms. These spiky creatures have 5 arms coming from a central disk. They are flexible and can move quickly along the sea floor. Starfish mainly eat clams and oysters. A starfish can grip a clam or oyster and pull on its shell until the creature tires. If the clam opens its shell only a fraction of an inch, the starfish will turn its own stomach inside out through its mouth. Forcing its stomach into the crack, the starfish then digests the clam while it is still in its shell. When it is done eating, the starfish pulls its stomach back inside its body and moves on.

Starfish also have the ability to regenerate. If one of its limbs is cut off it can grow another one.

Sea urchins, sand dollars, and starfish all begin as eggs, hatch into a larva and then grow into adults. Sea urchins have very long spikes. Sand dollars have very short spikes. Echinoderms are often very brightly colored.

Starfish Model:

Help your child use the salt dough to form a 5-legged starfish on a piece of tag board or cardboard. Encourage him to make the model thicker in the middle and thinner at the ends of the legs. Have him gently press mini-chocolate chips into the starfish to represent spikes. Take a picture of this model to include in your animal kingdom notebook.

Observing Starfish and Sand Dollars:

If possible, obtain a real (dead) starfish or sand dollar. These are sometimes available at craft or novelty shops. Observe each creature with a magnifying glass. Look for five legs and spiny skin. A sand dollar does not have five legs, but does have markings for five sections.

What did we learn?

What are three common echinoderms? (Starfish, sand dollars, and sea urchins)

What do echinoderms have in common? (They all have spiky skin, and most have 5 body parts radiating from a central disk.)

Taking it further

Why would oyster and clam fishermen not want starfish in their oyster and clam beds? (Starfish can eat up to a dozen clams or oysters at a time. This hurts the fishermen's business.)

What would happen if the fishermen caught and cut up the starfish? (The starfish would regenerate resulting in more starfish. This happened in one fishing village. The fishermen thought they were getting rid of the starfish by cutting them in half, but actually ended up making many more of them.)

What purpose might the spikes serve on echinoderms? (Most spikes are used for protection from predators.)

Lesson

Sponges

How much water can a sponge hold?

Supply list:

- Paper

- Tempera paints

- Synthetic sponges (and if possible, a real sea sponge)

- Scissors

One of the simplest multi-celled invertebrates is the sponge. Sponges attach themselves to the sea floor. They have tube-like bodies with no complex systems. It is believed that sponges do not even have nerve cells. What sponges do have is lots of holes.

Water flows into pores, which are small openings or holes in the sides of the sponge. Oxygen and microscopic organisms are removed from the water as it flows through the sponge. Then the water and any waste products are released through an opening on the top of the sponge.

Like starfish, sponges can regenerate. If even a small piece is cut off of a sponge it can grow into a new sponge. In fact, some sponge farmers grow sponges by cutting them up, attaching them to cement blocks and lowering the blocks into the sea.

Sponges are often found in the same areas as coral. When an area becomes too crowded a sponge may become aggressive and overgrow a colony of coral. Sponges are immune to the poison darts shot out by coral and can eventually overtake a coral colony.

For many years sea sponges were harvested for use as cleaning tools. However, synthetic sponges are now much more popular and real sponges are used less frequently.

Scientists originally thought sponges were plants because they do not move. But studies have shown that sponges do not produce their own food so they cannot be plants. Also, sponges can reproduce with eggs and the larvae do move around before anchoring themselves to the sea floor, thus classifying them as animals.

Sponge Painting:

Help your child cut synthetic sponges (the kind you get at the grocery store) into the shapes of jellyfish, coral, starfish, sand dollars, and other sea creatures. Help your child create an underwater picture by dipping each sponge into paint and pressing it on a piece of paper. When it is dry, this picture can be added to the animal kingdom notebook.

If a real sea sponge is available, examine it closely. Compare and contrast a real sea sponge with a synthetic sponge. How are they alike? How are they different? Which one would you prefer for cleaning your house?

What did we learn?

How does a sponge eat? (Nutrients are absorbed from the water as it passes through the body of the sponge.)

How does a sponge reproduce? (A sponge can reproduce by releasing eggs or a sponge can regenerate to form new sponges from pieces that are cut off of the original sponge.)

Why is a sponge an animal and not a plant? (A sponge cannot produce its own food and it reproduces with eggs so it is an animal.)

Taking it further

Why can a sponge kill a coral colony?(It is immune to the poison darts of the coral.)

What uses are there for sponges? (They are sometimes used for cleaning. But mostly they are used for sponge painting and other artwork.)

Why are synthetic sponges more popular than real sponges? (They are much less expensive.)

Worms

Creepy crawlers

Supply list:

- Gummy worms	Snack idea:
- Shoe box	- Instant chocolate pudding
- Rocks	- Crushed chocolate cookies
- Dirt, dried leaves	- Gummy worms

I f you are a fisherman then you probably know where to find worms. To many people earthworms are nothing more than fish bait. However, worms are much more important. There are three main groups of worms: segmented worms, flat worms, and round worms. All are long and narrow and have very simple bodies.

Segmented worms, worms with rings, are the most common. Nearly everyone is familiar with the earthworm. This creature loves moist earth. It eats dead plant material, turning it into fertilizer for plants to use. This is why worms are so important. Segmented worms are great composters. Many people raise worms to use in compost bins. You feed your food scraps to the worms and they turn it into compost or fertilizer for your garden. Sea worms, leeches, and rag worms are also segmented worms.

The second type of worm is the flat worm. As the name says, these are flat creatures. Most flat worms live in water or are parasites living inside animal hosts. Planarians are flat worms that live in water. They have arrow shaped heads and are usually less than one inch long. Planarians have a great ability to regenerate and if cut into pieces, all but the tail will grow into a new worm. Flukes and tapeworms are both parasitic flat worms. They survive by infesting a host animal and absorbing nutrients from it. Parasitic worms are very dangerous, and often deadly, to their hosts.

The third group of worms is round worms. These long, thin, smooth worms are almost all parasites. They often live in the intestines of the host and suck the host's blood or absorb digested food. They are almost always harmful to the host. So, depending on the type of worm, it can be very harmful or very beneficial to humans.

Worm Diorama:

Have your child make a scene in a shoebox showing an earthworm's habitat. Include dirt, rocks, dried leaves and any other items you might find where earthworms live. Use gummy worms for the earthworms. Take a picture of your diorama for your notebook.

Wormy Snack Idea:

Mix instant chocolate pudding according to package directions. Place about an inch of crushed chocolate cookie crumbs in the bottoms of 4 plastic cups. Put a gummy worm in each cup with one end hanging over the rim. Divide the pudding between the cups. Add another layer of cookie crumbs. Now you have 4 yummy mud pies with worms for dessert.

What did we learn?

What kinds of worms are beneficial to man? (Segmented worms such as earthworms)

How are they beneficial? (They break down dead plant material, and can be used as fishing bait.)

What kinds of worms are harmful? (Most other kinds of worms are parasitic and thus are harmful to their hosts, whether they are human or animal hosts.)

Taking it further

How can you avoid parasitic worms? (Parasitic worms thrive in unsanitary conditions and are much more of a threat in undeveloped countries. Washing hands and raw vegetables and cooking meat well will help you avoid most parasites.)

Invertebrates Quiz

Lessons 20-31

Answer True or False for each statement below.

1. _____ All invertebrates have backbones.

2. _____ Invertebrates have exoskeletons.

3. _____ Most invertebrates are small.

4. _____ An octopus is considered to be the most intelligent invertebrate.

5. _____ Bivalves have only one part to their shells.

6. _____ Insects have three body parts.

7. _____ All arachnids have eight legs.

8. _____ Grasshoppers experience incomplete metamorphosis.

9. _____ Some invertebrates have the ability to regenerate.

10. _____ All worms are harmful to people.

Give a short answer to each question below.

11. What do jellyfish, coral, and sea anemones have in common? _____

_____

12. What does a spider eat?_____

13. Why should scorpions be avoided? _____

14. How can insects be useful? _____

15. Why should you check for ticks after hiking through the woods? _____

_____

(Note: answers to all quizzes and tests are in Appendix A)

Protists

Single-celled creatures

Supply list:

- Construction paper	- Yarn
- Shoe	- Scissors
- Colored pencils, markers, or crayons	- Glue

The simplest life forms are the single-celled protists. However, these microscopic creatures are more complex than you might think. Protists have all of the basic parts of an animal cell including cell membrane, nucleus, cytoplasm, mitochondria, and vacuoles. The cell membrane acts like skin – providing protection. The nucleus acts like the brain and controls the cell's functions. The cytoplasm provides a transportation network for the various parts of the cell. The mitochondria are the cell's power plants. They break down food and provide energy. And the vacuoles are the cell's warehouses – providing food storage.

In addition, most protists have specialized parts that allow them to perform many of the functions that larger creatures do. They eat and digest food, breathe, move, and protect themselves. There are thousands of protists. Scientists have grouped them by the way they move.

Flagellates are single-celled creatures that move by using a flagellum or whip-like structure at the front of the cell. A euglena is a common flagellate found in freshwater lakes and ponds. It uses its flagellum like an outboard motor to propel itself through the water.

The euglena is a puzzling creature because, even though it has the characteristics of an animal and has the ability to catch food, it also has chlorophyll in its body and can produce its own food. Because of this and other anomalies, protists are usually put into a kingdom of their own and are not necessarily grouped with animals.

The second type of protist is the sarcodine. These are single celled-creatures with a pseudopod. Pseudopod means false foot. A sarcodine moves by extending one part of its cell membrane in a finger or foot-like projection and then moving the rest of the cell into that area. The amoeba is the most familiar creature with pseudopods. It is continually moving by changing its shape. An amoeba generally has several pseudopods sticking out at any one time. An amoeba ingests food by extending two or more pseudopods to surround the food and thus take it into its cell.

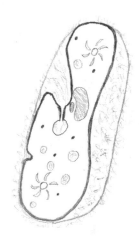

The third type of protist is the ciliate. Ciliates are single-celled animals that are surrounded by cilia or hair-like projections that propel them through the water. A paramecium is a common ciliate. A paramecium is a submarine-shaped cell. The cross-section of a paramecium shown here displays the internal parts of the cell. The actual paramecium is covered with cilia on all surfaces. Its cilia not only move it around but also push food into its gullet (an opening that serves as its mouth).

Most protists live in water. Many are parasitic and cause some very serious diseases such as malaria, African sleeping sickness, and amoebic dysentery. Protists generally reproduce by some sort of division where one cell divides to form two new cells. Even though these are the simplest life forms, God's amazing design is still very obvious in their complex functions.

Paramecium Model:

Have your child make a model of a paramecium. Have him trace his shoe on a piece of construction paper then cut it out. He can glue short pieces of yarn around the edges to represent the cilia. He can cut different colors of paper to represent the nucleus, vacuoles, and mitochondria and glue them to the model. Note: this is a two dimensional or flat model. Actual paramecium are more submarine shaped and covered all over with cilia. Add this model to your animal kingdom notebook.

Optional – Observe Microscopic Creatures:

If you have a microscope available, examine a drop of pond or stream water. Look for tiny creatures that live in the water. You may be able to observe some of the creatures discussed in the lesson as well as slightly larger creatures.

What did we learn?

How are protists different from other animals? (They consist of only one cell. Some contain chlorophyll.)

How are they the same? (Protists reproduce, eat, move, grow, and need oxygen just like other animals. Also, protists have all the same cell parts as other animal cells.)

Taking it further

Why is a euglena a puzzle to scientists? (It has plant and animal characteristics.)

Why are single-celled creatures not as simple as you might expect? (Just because there is only one cell does not mean it is simple. Single-celled creatures perform very complex functions. Most protists are more complex than any cell in the human body because they cannot be as specialized. Even the smallest organism demonstrates God's marvelous powers of design.)

Lesson

33

Monerans

Bacteria and viruses

Supply list:

- Hand soap and other anti-bacterial items in your house

The final group of living organism is the monerans. These creatures are the germs that make you sick as well as the organisms that help recycle minerals from dead plants and animals. These creatures fit into two groups: bacteria and viruses. Monerans are usually not classified as animals or plants. They are usually classified in their own kingdom.

Bacteria are single-celled creatures. However, they do not have a defined nucleus like protists do. Some bacteria can produce their own food while others feed off other cells or dead plants and animals. Some bacteria make humans sick. Bacteria can cause plague, pneumonia, and tuberculosis. But not all bacteria are harmful. Most bacteria are very helpful. Bacteria are vital in the breakdown of dead plants and animals. Also, bacteria are necessary in the human digestive system. Without bacteria, our bodies cannot properly digest the food we eat.

Viruses are some of the smallest "creatures" yet they present some of the biggest puzzles to scientists. Scientists are not sure if viruses are even alive. A virus has genetic information like a cell but it does not directly reproduce. Instead, it invades a host cell and reprograms it to reproduce more viruses. Most of the diseases we are familiar with are caused by viruses, including flu, the common cold, chicken pox, and measles.

Scientists have discovered antibiotics that can be used in the treatment of bacteria induced diseases. However, few treatments have been found to cure diseases caused by viruses. Many of the more serious diseases can be prevented by the use of vaccines, which encourage your body to build up defenses against certain viruses but cannot cure the diseases.

Anti-bacterial Hunt:

Many of the items in our homes are anti-bacterial; that is, they kill bacteria. Search your house looking at things like hand soap, laundry soap and cleaning supplies to see how many of them say "anti-bacterial" on them. Also check medical supplies such as anti-bacterial creams or sprays, bandages, etc. People have become concerned about germs and want products that get rid of them in hopes of staying healthier.

What did we learn?

How are bacteria similar to plants and animals? (They have cells, reproduce, and some can produce their own food.)

How are bacteria different from plants and animals? (They do not have a defined nucleus.)

How are viruses similar to plants and animals? (They have genetic information--DNA.)

How are viruses different? (They do not reproduce on their own. They do not eat or grow in a normal sort of way.)

Taking it further

Ask the following questions to test if a virus is alive. Does it have cells? (Not sure) Can it reproduce? (Only with the help of a host cell) Is it growing? (Can't tell, they are too small to see even with an electron microscope) Does it move or respond to its environment? (Yes) Does it need food and water? (It needs host cells that use food and water. It is unclear if the viruses use these things directly.) Does it have respiration? (Also unknown) Is it alive? (We aren't sure.)

How can use of antibiotics be bad? (Antibiotics kill bacteria, but they cannot distinguish between good and bad bacteria. Overuse of antibiotics can kill too many of the good bacteria in your intestines and cause problems. Also, antibiotics kill most of the bad bacteria but some are resistant and do not die. These bacteria are the ones that survive and reproduce. The next generation of bacteria is not as easily killed by the antibiotics. Doctors are beginning to see diseases that used to respond to certain antibiotics no longer respond and must be treated with stronger medicines. So we need to carefully use antibiotics when necessary, but not overuse them or use them incorrectly.)

Got Milk?
Louis Pasteur
(1822-1895)

If you've got milk you might want to thank a French chemist named Louis Pasteur who was born two days after Christmas in 1822. His father had served in Napoleon's Army and afterward worked as a tanner.

Louis Pasteur did a lot of work that we are still thankful for today, like what he did for milk. He came up with a way of processing the milk to kill off the bacteria so it will stay good for more than a couple of days. The process was named after him. We call it pasteurization (look on any milk container).

He helped us in many other ways, too. Today, a woman can go to the hospital to deliver a baby and is able to enjoy the gift of a new life coming into the world without the fear of dying from infection. In Louis Pasteur's day, about 1/3 of the pregnant women in Paris died from childbirth fever or infection. Pasteur convinced the medical community that their sloppy practices were spreading germs and hurting their patients.

However, most of his ideas were not accepted easily. When he said doctors should wash their hands and sterilize their instruments he really upset the medical community. He was called a menace to science. They said, "Who does Pasteur think that he is? He isn't even a medical doctor...just a lowly chemist."

The wife of the Emperor asked Dr. Pasteur to come explain his radical views to the French Court. He told the Emperor that the hospitals in Paris were death houses and that most of the doctors carried death on their hands (referring to germs). When he accurately predicted the death of the Emperor's sister-in-law he was condemned as a fraud and banned by the Emperor from speaking out in public about medicine.

After this, Pasteur moved to the countryside where he spent the next ten years working to discover the causes of anthrax, the black plague of sheep. Anthrax had been ravaging the sheep across France. Pasteur invented an anthrax vaccine, which he gave to the farmers to use on their sheep for free.

At this time the French government needed more sheep to pay the 5 million francs they owed to Germany for their war indemnity. They came to the area where Pasteur had been working with the farmers to find out why their sheep were so healthy. When Pasteur told them of his vaccine he was again mocked as a fool by the Academy of Medicine. He showed them the truth by taking 50 sheep and vaccinating 25 of them. Then all 50 were infected with blood carrying anthrax. To everyone's amazement, only the sheep that had been vaccinated survived. Because of Pasteur's work, we now have a reliable cure for anthrax for both livestock and humans.

Even with this wonderful success, the medical establishment was slow to accept Pasteur. Eventually however, he was elected as a member of the Académie Française in 1882. There he undertook the task of finding a cure for rabies. Three years later he was able to save the life of a young boy named Joseph Meister who had been bitten by a rabid dog. The boy survived and later become the caretaker of Pasteur's tomb at the world-famous Pasteur Institute in Paris. Louis Pasteur headed work at the Pasteur Institute, which was inaugurated in Paris in 1888, until his death on September 28, 1895.

Dr. Pasteur's work has saved millions of lives, but his discoveries came too late to save three of his daughters, who died from typhoid fever. Pasteur selflessly taught that the benefits of science are for all of humanity, not for the benefit of the scientist, and today all of humanity is reaping the benefits of his work.

Lesson

34

Animal Kingdom Notebook

Putting the animals together

Unit Project supply list:

- Paper

- Art supplies

- Clip art or other animal pictures

- Photographs of projects from previous lessons

- Worksheets from previous lessons

After learning about all the different animals that God created we can see that He created a wonderful world of life. From the most complex vertebrate to the simplest single-celled organism we can see the hand of the Master Creator. You have been making a notebook with all of your projects from this book. Now take what you have learned and finish up your book so you can share the animal kingdom with someone else.

Unit Project

Animal Kingdom Notebook:

As you have been studying the animal kingdom, your child has been building a notebook. Now it is time to complete the notebook. You can make this as simple or complex as you desire. Some ideas are given below, but feel free to make whatever changes suit what you want your child to learn.

It will be very beneficial to have library books available to provide additional information and ideas for the pages of your notebook.

It will probably take several days to complete this book. When you are done you want to have something that your child can be proud of.

Some ideas for making pages in your notebook:
- Older children can write a report for each section of the book.
- If using a computer, add clip-art to your pages. Many pictures of animals are available in clip-art files.
- Be creative. Don't require that every section "look" like every other section.
- Clip pictures from old magazines or coloring books to add to your book.
- If your child is artistic he can draw pictures of many animals.
- Take photographs of projects your child has completed in previous lessons and include them in the notebook.
- Add photographs of field trips you have taken.
- Include the worksheets your child has completed in previous lessons.

Title Page: Have your child make a title page.

Table of Contents: This will allow readers to find information quickly.

Vertebrates Section: Should include information for all 5 types of vertebrates.
Mammals
Birds
Amphibians
Reptiles
Fish

Invertebrates Section: Should include information for all 6 types of invertebrates.
    Arthropods
    Mollusks
    Coelenterates
    Echinoderms
    Sponges
    Worms

Protists and Monerans Section: Include information for these microscopic creatures.
    Flagellates
    Sarcodines/Pseudopods
    Ciliates
    Bacteria
    Viruses

What did we learn?

What do all animals have in common? (They are alive, they reproduce, they do not make their own food, they can move about during at least part of their life.)

What is the difference between vertebrates and invertebrates? (Vertebrates have a backbone and invertebrates do not.)

What sets protists apart from all the other animals? (They are single-celled creatures. Some, like the euglena, can make their own food.)

Taking it further

What are some of the greatest or most interesting things you learned from your study of the animal kingdom?

Read Genesis chapters 1 and 2. Discuss what was created on each day and how each part completes the whole.

What would you like to learn more about? (Check out books from the library to learn more.)

Animal Kingdom Unit Test

Lessons 1-34

Match each animal group with its unique characteristic.

1. _____ Mammals   A. Fins and gills

2. _____ Birds   B. Jointed feet or jointed legs

3. _____ Fish   C. Mantle that produces a shell

4. _____ Reptiles   D. Nurses its young

5. _____ Amphibians   E. Single-celled

6. _____ Arthropods   F. Dry scaly skin

7. _____ Mollusks   G. Feathers

8. _____ Echinoderm   H. Gills and lungs

9. _____ Coelenterates   I. Spiny skin

10. _____ Protists   J. Hollow body

Define the following terms:

11. Invertebrate: _____

12. Vertebrate: _____

13. Cold-blooded animal: _____

14. Warm-blooded animal: _____

15. Moneran: _____

Describe how God designed a bird's feet for each task listed below:

16. Swimming in a lake: _____

17. Perching in a tree: _____

18. Hunting prey: _____

19. Describe how a bird is specially designed for flight. _____

_____

_____

20. Name the three body parts of an insect.

_____   _____   _____

21. Name the two body parts of a spider.

_____   _____

Answer True or False for each statement below:

22. _____ Snakes have a special organ for sensing smell.

23. _____ Cold-blooded animals do not need to eat as often as warm-blooded animals.

24. _____ Turtles can safely be removed from their shells.

25. _____ Cartilaginous fish do not have any bones.

26. _____ Centipedes are arthropods.

27. _____ All crustaceans live in the water.

28. _____ Insects are the most common arthropod.

29. _____ Some creatures can live closely with jellyfish.

30. _____ The best way to kill a starfish is to cut it in half.

(Note: answers to all quizzes and tests are in Appendix A)

Conclusion

Reflecting on the animal kingdom

Supply list:

> - Bible
>
> - Paper and pencil

We have studied the animal kingdom with its nearly infinite variety of creatures. As we think about the world of animals around us we should be thankful to God for the wonderful job He did. Take a few minutes and contemplate how glorious God's creation is and realize that it is indeed GOOD!

Thanksgiving to God:

Read Job 38:39-40:5. Discuss all the wonders mentioned in this passage. Discuss Job's response to God's questions. How should we respond to God's creation?

Write a poem or prayer of thanksgiving to God for the world of animals.

Appendix A
Answers to Quizzes and Test

Mammals Quiz Answers:

1. vertebrates and invertebrates   2. mammals, birds, amphibians, reptiles, fish   3. warm-blooded, breathes air with lungs, fur/hair, live birth, nurses young   4. has a pouch   5. has a spinal cord ending in a brain   6. F   7. F   8. T   9. F   10. F Elephants are the largest land animal, but blue whales are the largest animal on earth.   11. T   12. T   13. T   14. T   15. F

Vertebrates Quiz Answers:

Vertebrates have backbones with a spine along the back ending in a brain.

Mammals: A, C, G, J, L, (H for some)
Birds: A, E, H, J, M
Fish: B, D, H, K N, (G for a few)
Reptiles: B, D, H, J
Amphibians: B, F, H I , J, K

Arthropods Quiz Answers:

1. Y   2. Y   3. N   4. Y   5. Y   6. Y   7. Y   8. Y   9. N   10. N   11. Y   12. N   13. Y   14. Y   15. N   16. N

17. egg, larva, pupa (or chrysalis), adult   18. 3, 6, 2, 2 or 4   19. 2, 8, 0, 0   20. 1, 2

Invertebrates Quiz Answers:

1. F   2. T   3. T   4. T   5. F   6. T   7. T   8. T   9. T   10. F

11. They have hollow bodies and stinging tentacles.   12. It sucks liquids from the bodies of other animals.

13. They are poisonous.   14. Pollination of plants, eat other pests, provide food for other animals.

15. Ticks often live in the woods. If they attach themselves to you they can suck your blood and transmit diseases.

Animal Kingdom Unit Test Answers:

1. D   2. G   3. A   4. F   5. H   6. B   7. C   8. I   9. J   10. E

11. Invertebrates are creatures without a backbone.

12. Vertebrates are creatures with a backbone.

13. Cold-blooded animals cannot regulate their body temperature; it is the same as the surrounding temperature.

14. Warm-blooded animals regulate their body temperature to keep it the same regardless of the surrounding temperature.

15. Monerans are bacteria and viruses. They are tiny creatures that do not fit into any other category.

16. Swimming: webbed feet   17. Perching: three toes facing forward, 1 toe facing backward for grasping tree branches.   18. Hunting: sharp claws (or talons) for grasping prey.

19. Possible answers include: Birds have hollow bones, air foil shaped wings, contour feathers that point toward the back of the body, special flight feathers, a tail that works like a rudder, very efficient respiratory and circulatory systems.

20. Head, thorax, abdomen   21. Cephalothorax, abdomen

22. T   23. T   24. F   25. T   26. T   27. F   28. T   29. T   30. F

Appendix B
Resource Guide

Suggested Library Books

Reader's Digest North American Wildlife – Our favorite resource to have for any field trip!
Usborne's Illustrated Encyclopedia of the Natural World – Great overview of most animals
Simon and Shuster's Young Readers' Book of Animals by Martin Walters – Excellent overview of the animal kingdom
Biology for Every Kid by Janice VanCleave – Many fun activities
National Geographic Book of Mammals – Great pictures – A to Z listings
Mammals Scholastic Voyages of Discovery by Scholastic Books –Interactive fun
Birds by Carolyn Boulton – Lots of suggested activities, good pictures
What is a Fish? by Bobbie Kalman and Allison Larin – Good overall discuss of fish
Play and Find Out About Bugs by Janice VanCleave – Great experiments
Jellyfish by Leighton Taylor – Good explanation of lifecycle, great pictures

Suggested Videos

Newton's Workshop by Moody Institute – Excellent Christian science series

Science Supply Resources

Home Training Tools
1-800-860-6272
www.hometrainingtools.com

Tobin's Lab
1-800-522-4776
mike@tobinlab.com

The Science Resource
972-644-4452
www.thescienceresource.com

Blue Spruce Biological Supply
1-800-825-8522
http://bluelio.com

We highly recommend purchasing one or more of the following from one of the above suppliers to supplement the activities in this book:
Owl Pellets   Grow-a-frog   Butterfly habitat   Dissection supplies

Creation Science Resources

The Answers Book by Ken Ham and others – Answers frequently asked questions
Dinosaurs by Design by Duane T. Gish – All about dinosaurs and where they fit into creation
The Amazing Story of Creation by Duane T. Gish – Gives scientific evidence for the creation story
Creation Science by Felice Gerwitz and Juill Whitlock – Unit study focusing on creation
Creation Facts of Life by Gary Parker – In depth comparison of the evidence for creation and evolution

Field Trip Ideas

Farm or dairy
Zoo, aquarium, or butterfly museum
Fish hatchery
Wildlife area

Appendix C
Master Supply List

Supplies needed	Lesson
3-ring binder	2
Dividers with tabs (12 or 13 per child)	2
Hair/fur from 2 or more mammals	3
Toothbrush	6
Stop watch	6
Chopped fruits, nuts, or vegetables	6
Fake fur or felt	7
Plastic zipper bag	7
Tag board or card board	7, 29
Magnifying glass	9, 25
Bird feather (can purchase at craft store)	9
Sequins or flat beads	13
Face paint	15
Cloth tape	16
Goldfish snack crackers	17
Modeling clay	19, 25, 28
Styrofoam balls (3 per child)	22
Pipe cleaners	22, 24
Toothpicks	22, 24
Index cards	22
Sleeping bag	23
2 scarves	23
Marshmallows (large and small)	24, 28
Flexible wire	24
Several sea shells	27
Sea shell identification guide	27
Mini-chocolate chips	29
Real starfish (dead – check at craft store)	29
Real sand dollar (dead – check at craft store)	29
Sponges (synthetic kind you get at grocery store)	30
Tempera Paints	30
Real sea sponge (check at home decorating store)	30
Gummy worms	31
Shoe box	31
Yarn	32
Anti-bacterial hand soap	33

Optional Supplies:	Lesson
Owl pellet	10
Tadpoles and tank	12
Butterfly larvae (caterpillars)	23

Appendix D
List of Reproducible Pages

Permission is granted to reproduce the pages listed below for single classroom use.

Index

Works Cited

Adams, A. B. *Eternal Quest: The Story of the Great Naturalists*. New York: G.P. Putnam's Sons, 1969.

Bargar, Sherie, and Linda Johnson. *Rattlesnakes*. Vero Beach: Rourke Enterprises, Inc., 1986.

Cardwardine, Mark, et.al. *Whales, Dolphins & Porpoises*. Sydney: US Weldon Owen Inc., 1998.

Chinery, Michael. *Butterfly*. Mahwah: Troll Associates, 1991.

Chinery, Michael. *Shark*. Mahwah: Troll Associates, 1991.

Coldrey, Jennifer. *Shells*. New York: Dorling Kindersley, Inc., 1993.

Cole, Joanna. *A Bird's Body*. New York: William Morrow & Co., 1982.

Cousteau Society. *Corals The Sea's Great Builders*. New York: Simon & Shuster, 1992.

'Espinasse, M. *Robert Hooke*. Berkeley: University of California, 1962.

Evans, J. Edward. *Charles Darwin Revolutionary Biologist*. Minneapolis: Lerner Publications, 1993.

Fleisher, Paul. *Gorillas*. New York: Benchmark Books, 2001.

Gish, Duane T., Ph.D. *The Amazing Story of Creation*. El Cajon: Institute for Creation Research, 1990.

Gowell, Elizabeth Tayntor. *Whales and Dolphins What They Have in Common*. New York: Franklin Watts, 1999.

Ham, Ken. *The Great Dinosaur Mystery Solved!* Green Forest: Master Books, 1999.

Hird, Ed. "Dr. Louis Pasteur: Servant of All." *Deep Cove Crier*. December 1997: n.pag.

http://web.ukonline.co.uk. *Louis Pasteur*. n.p.: n.p., n.d.

http://www.earthlife.net/birds. *Birds*. n.p.: Earth-Life Web Productions, 2002.

http://www.ucmp.berkeley.edu/history/cuvier. *Georges Cuvier*. Berkeley: Univeristy of California, 2002.

Jackson, Tom. *Nature's Children Rattlesnakes*. Danbury: Grolier Educational, 2001.

Kalman, Bobbie, and Allison Larin. *What is a Fish?* New York: Crabtree Publishers, 1999.

Koerner, L. *Linnaeus: Nature and Nation*. Cambridge: Harvard University, 1999.

Lacey, Elizabeth A. *The Complete Frog a Guide for the Very Young Naturalist*. New York: Lothrop, Lee & Shepard Books, 1989.

Landau, Elaine. *Sea Horses*. New York: Children's Press, 1999.

Lindroth, S. *The Two Faces of Linnaeus*. Berkeley: University of California Press, 1983.

Maynard, Thane. *Primates Apes, Monkeys, Prosimians*. New York: Franklin Watts, 1994.

Markle, Sandra. Outside and Inside Kangaroos. New York: Atheneum Books for Yound Readers, 1999.

Moore, J. A. *Science as a Way of Knowing*. Cambridge: Harvard University Press, 1993.

Morris, John D., Ph.D. *The Young Earth*. Colorado Springs: Master Books, 1994.

National Geographic Book of Mammals. Washington, D.C.: National Geographic Society, 1998.

Parker, Gregory, and others. *Biology God's Living Creation*. Pensacola: A Beka Books, 1997.

Parker, Steve. *Charles Darwin and Evolution*. London: HarperCollins Publishers, 1992.

Ross, Michael E. *Wormology*. Minneapolis: Carolrhoda Books, Inc., 1996.

Rudwick, M.J. S. *The Meaning of Fossils*. Chicago: University of Chicago Press, 1985.

Scholastic Voyages of Discovery Mammals. New York: Scholastic, Inc., 1997.

Stone, Lynn M. *Tasmanian Devil*. Vero Beach: Rourke Corptoration, Inc., 1990.

Swan, Erin Pembrey. *Meat-eating Marsupials*. New York: Franklin Watts, 2002.

Swan, Erin Pembrey. *Primates From Howler Monkeys to Humans*. New York: Franklin Watts, 1998.

Taylor, Leighton. *Jellyfish*. Minneapolis: Lerner Publishing Co., 1998.

VanCleave, Janice. *Biology for Every Kid*. New York: John Wiley & Sons, Inc., 1990.

VanCleave, Janice. *Insects and Spiders*. New York: John Wiley & Sons, Inc., 1998.

VanCleave, Janice. *Play and Find Out About Bugs*. New York: John Wiley & Sons, Inc., 1999.

Walters, Martin. *The Simon & Schuster Young Readers' Book of Animals*. New York: Simon & Schuster, Inc., 1990.